Research Topics in Wind Energy

Volume 6

About this Series

The series Research Topics in Wind Energy publishes new developments and advances in the fields of Wind Energy Research and Technology, rapidly and informally but with a high quality. Wind Energy is a new emerging research field characterized by a high degree of interdisciplinarity. The intent is to cover all the technical contents, applications, and multidisciplinary aspects of Wind Energy, embedded in the fields of Mechanical and Electrical Engineering, Physics, Turbulence, Energy Technology, Control, Meteorology and Long-Term Wind Forecasts, Wind Turbine Technology, System Integration and Energy Economics, as well as the methodologies behind them. Within the scope of the series are monographs, lecture notes, selected contributions from specialized conferences and workshops, as well as selected PhD theses. Of particular value to both the contributors and the readership are the short publication timeframe and the worldwide distribution, which enable both wide and rapid dissemination of research output. The series is promoted under the auspices of the European Academy of Wind Energy.

More information about this series at http://www.springer.com/series/11859

Gijs van Kuik · Joachim Peinke
Editors

Long-term Research Challenges in Wind Energy - A Research Agenda by the European Academy of Wind Energy

Springer

Editors
Gijs van Kuik
Wind Energy Research Institute (DUWIND)
Delft University of Technology
Delft
The Netherlands

Joachim Peinke
Institute of Physics
University of Oldenburg
Oldenburg
Germany

and

Fraunhofer - IWES Institute for Wind Energy and Energy Systems Technology
University of Oldenburg
Oldenburg
Germany

ISSN 2196-7806 ISSN 2196-7814 (electronic)
Research Topics in Wind Energy
ISBN 978-3-319-46918-8 ISBN 978-3-319-46919-5 (eBook)
DOI 10.1007/978-3-319-46919-5

Library of Congress Control Number: 2016951975

Printed on acid-free paper

This Springer imprint is published by Springer Nature
The registered company is Springer International Publishing AG
The registered company address is: Gewerbestrasse 11, 6330 Cham, Switzerland

Preface

The European Academy of Wind Energy (eawe, http://www.eawe.eu), representing universities and institutes with a significant wind energy programme in 14 countries, has discussed the long-term research challenges in wind energy. In contrast to research agendas addressing short- to medium-term research activities, this eawe document for the first time puts its focus explicitly on a longer-term perspective, addressing the scientific knowledge base that is required to develop wind energy beyond the applications of today and tomorrow. In other words, this long-term research agenda is driven by problems and curiosity, addressing basic research and fundamental knowledge in 11 research areas, ranging from physics and design to environmental and societal aspects. Our basic claims are that the relatively young developments in wind energy as a new renewable technology needs the input from basic sciences to advance further, as well as that highly challenging scientific questions come up with advancements of this technology. Because of the very nature of this initiative, this document does not intend to be permanent or complete. It shows the vision of the experts of the eawe, but other views may be possible. We sincerely hope that it will spur an even more intensive discussion worldwide within the wind energy community. To enable this a discussion forum has been opened by the eawe at http://www.eawe.eu/index.php/research/forum-agenda. The reader is invited to contribute to this discussion.

The content of this book, without pictures and photos, has been published as the very first paper in the open-access journal Wind Energy Science launched in 2016 by the eawe and Copernicus Publications, see http://www.wind-energ-sci.net/1/1/2016/.

Delft, The Netherlands — Gijs van Kuik
Oldenburg, Germany — Joachim Peinke

Acknowledgements

Thanks go to the European Academy of Wind Energy, eawe, who offered the platform for the discussion on the R&D challenges. The eawe has once again shown that it is thé organisation in Europe for the acceleration and promotion of scientific research in wind energy, with the best researchers joining forces in it.

Also the Hanse–Wissenschaftskolleg (Institute for Advanced Study, HWK) is gratefully acknowledged. This institute granted a research fellowship to the first editor, resulting in very good cooperation on this "research challenges" project. HWK organised and hosted the round-table discussions. The stimulating atmosphere of the institute and the efficient workshop organisation made the meetings a productive and enjoyable experience for all participants. More information on HWK is given at the last page.

We thank all experts who contributed to this document by attending the round-table discussions at the HWK or by commenting on draft texts. We are very grateful for all discussions on the research challenges that have taken place in many institutes and helped to formulate this document. We mention in particular, and hopefully not forgetting any names, Sandrine Aubrun, Jens Bange, Stefan Barth, Dennis Bosse, Geoff Dutton, Peter Eecen, Ben Hendriks, Athanasios Kolios, William Leithead, Christian Liewen, Mart van der Meijden, Roland Ortt, Andreas Rettenmeier, Raimund Rolfes, Daniel Schertzer, Elke Seidel, Wolfgang Stenzel, Eize de Vries, Laurens de Vries, Matthias Wächter, Simon Watson, and Jonathan Whale.

Finally, thanks go to Jos Beurskens, SET Analysis, Schagen NL, who has provided most of the photos in the book, selected from his huge collection, and to several others who delivered illustrations, see page VIII.

Copyright and Justification of the Figures:

Number	Copyright	Place
0.1	Elke Seidel,	University Oldenburg
0.2	Jos Beurskens	
0.3	Jos Beurskens	Bremerhaven
1.1	Jos Beurskens	Bremerhaven
1.2	Jos Beurskens	Bremerhaven
2.1	Jos Beurskens	Northern Germany
2.2	Jos Beurskens	Waddinxveen
3.1	Dominik Traphan	University Oldenburg
3.2	Jos Beurskens	Wieringermeer
4.1	Edwin van Solingen	Delft University of Technology
4.2	Jan-Willem van Wingerden	Delft University of Technology
5.1	Jos Beurskens	GE, Salzbergen
5.2	Jos Beurskens	GE, Salzbergen
6.1	Jos Beurskens	Wieringermeer
6.2	Jos Beurskens	Falck, Rotterdam
7.1	ECN	Petten
7.2	Jos Beurskens	Bremerhaven
8.1	MARIN	Wageningen
8.2	Franciska Klein	Off the coast of Aguçadouro, Portugal
9.1	Jos Beurskens	Dutch North Sea
9.2	Jos Beurskens	Off the coast of North Holland
10.1	Jos Beurskens	Bremerhaven
10.2	Jos Beurskens	Bremerhaven
11.1	Jos Beurskens	Highway in northern Germany
11.2	Jos Beurskens	Village in central Germany

Contents

Introduction

Energy is one of the central and foremost grand challenges facing our society today. To achieve a sustainable energy supply, solar and wind energy will likely provide the main renewable contributions to the growing need for clean energy for at least the next few decades. For solar energy, and in particular for photovoltaics, it is commonly accepted that there is a need for basic fundamental research. With this document, we want to proclaim that there is also a strong need for basic long-term research in wind energy.

The European Academy of Wind Energy (eawe) presents here its view on the long-term research challenges that will need to be overcome for the future development of wind energy. This overview addresses fundamental and pioneering research questions that, if solved, would have a significant impact on the understanding of the wind energy conversion process and that, in turn, would accelerate the development and penetration of wind in the energy mix.

In many countries, a discussion is ongoing to prioritise topics for wind energy research. National and international funding and scientific organisations often consider wind energy to be an engineering discipline that does not involve fundamental, basic or pioneering research. It is time for the international wind energy community to take a position in this discussion, and to show needs and benefits of a long-term scientific agenda. The eawe, which represents almost all European universities with a significant wind energy programme, is now willing to take responsibility for this action.

The main objectives of formulating the present long-term research agenda are as follows:

- To identify current technological and scientific barriers, stimulating new creative ideas to overcome them
- To define priorities for future scientific research
- To rethink our scientific view of wind energy
- To stimulate the cooperation among researchers in fundamental and applied sciences towards wind energy research

The eawe research agenda differs from the ones published by the International Energy Agency[1], the European Energy Research Alliance[2], the European Wind Energy Technology Platform[3] and other organisations. In fact, all of these tend to accelerate R&D and to promote technological developments that lower the cost of energy from wind on a short-time horizon. Consequently, the scope of work of those agendas is more focused on applied research than on fundamental or pioneering one. On the other hand, the eawe research agenda takes a longer-term perspective, addressing the scientific knowledge base that is required to develop wind energy beyond the applications of today and tomorrow. This agenda looks boldly forward towards scientific and technological breakthroughs or major advancements, without being constrained (yet) by short-time horizons driven by the need for rapid application, by fast returns in terms of a reduced cost of energy, or by immediate industrial relevance. In other words, this long-term research agenda is driven by problems and curiosity, addressing basic research and fundamental knowledge.

The present version of this document is the result of intensive discussions among many experts who have tried to describe the needs for basic long-term research to advance wind energy science and knowledge. At first, the discussions took place at several round-table talks organised by the Hanse-Wissenschaftskolleg (Institute for Advanced Study, HWK, see the last page) Delmenhorst in collaboration with the eawe, which led to the classification of the main general topics within the following specific items:

1. Materials and structures
2. Wind and turbulence
3. Aerodynamics
4. Control and system identification
5. Electricity conversion
6. Reliability and uncertainty modelling
7. Design methods
8. Hydrodynamics, soil characteristics and floating turbines
9. Offshore environmental aspects
10. Wind energy in the electric power system
11. Societal and economic aspects of wind energy

This initial classification work was then followed by open Internet discussions led by the main contributing authors of the various topics. We express our sincere appreciation to all contributors who have devoted time and effort to this goal and shared their knowledge and vision. As this document is the result of expert

[1]IEA Technology Roadmap Wind Energy 2013–2050 http://www.iea.org/publications/freepublications/publication/technology-roadmap-wind-energy—2013-edition.html

[2]EERA JP Wind Strategy for 2014–2030, http://www.eera-set.eu/eera-joint-programmes-jps/wind-energy/

[3]Strategic Research Agenda TP-Wind Energy, March 2014, http://www.windplatform.eu/fileadmin/ewetp_docs/Documents/reports/TPWind_SRA.pdf

discussions and is devoted to long-term research questions for the future, citations to the existing literature and to the state of the art were not included in the text.

Because of the very nature of this initiative, this document does not intend to be permanent nor complete. On the contrary, we sincerely hope that it will spur an even more intensive discussion worldwide within the wind energy community. Furthermore, we hope that it will strengthen the cooperation with experts in science and technology outside wind energy, as many of the research questions, although formulated from a wind energy perspective, will show links and overlap with their questions (Fig. 1).

The eawe intends to revise this document every 5 years. The international wind energy community is invited to join the discussion, for which a platform is opened at http://www.eawe.eu. Hopefully, not everyone will agree on everything that has been written in the detailed contributions appearing in the following and will consequently feel motivated to contribute to a future edition with their own ideas. We sincerely hope that this document shows that fundamental and long-term research contributes to the perspectives of wind energy and that it inspires all readers to satisfy their scientific curiosity and pursue their drive for knowledge, to the benefit of wind energy science.

Science and Wind Energy

As an introduction to the research challenge description in the various areas, some aspects of wind energy research that go beyond single research disciplines are discussed, starting with a brief history of modern wind energy development.

From a 10 m Diameter, 10 kW Turbine to a 1 GW Wind Power Station with 200 m Diameter, 15 MW Turbines

Modern wind energy started 40 years ago, driven by the first global oil crisis. In many countries governments started wind energy technology and demonstration programmes. As the general impression was that it should not be that difficult to construct large wind turbines given the knowledge accumulated in the aerospace and construction communities, all but one government asked their aerospace and construction companies to develop large wind turbines. During the 1980s, wind turbines in the range of 1 to 5 MW were built and tested by Boeing, Hamilton Standard, MBB, MAN, General Electric, Fokker and further companies. None of these machines proved to be successful, not to mention a reasonable cost of energy. All turbine designs were based on the best conceptual design solutions for rotating machinery like helicopters, including advanced pitch control, flexible and hinged blades, and variable speed. The aerospace route failed due to a severe underestimation of the wind turbine specific technology and its design drivers. It became clear that constructing an efficient wind turbine needs much more challenging work than just modifying a propeller or helicopter.

One country took another approach: Denmark stimulated the development of small turbines by pioneering individuals and small companies. After the failure of the aerospace route most other countries adopted this pioneer route too. This development, starting with wind turbines of 10 kW, slowly but surely led to the wind turbines of today (Fig. 2). The size of modern turbines as well as most of the early aerospace-driven concepts has been materialised in successful machines, operating on- and offshore. Wind turbines have become the largest rotating machines on earth, while plans for future turbines show even larger diameters up to 200 m, matching the dream of the late 1980s.

When wind turbines went to sea, a new but similar phase of underestimation of the wind turbine specific offshore knowledge started. The (oil and gas) offshore community had to learn that a wind turbine is not just another offshore device; likewise the onshore wind energy community had to learn by trial and error that onshore turbines are not good enough to survive at sea.

Having set the step from onshore to offshore, the step from wind turbine to wind power station was made at the same time. Wind power plants of 1 GW installed power have already been realised. In several countries wind power supplies a significant amount of the electricity consumption, so the interaction between wind power and the grid poses new challenges to the wind power stations as well as the grid.

Fig. 1 Scientists during the EAWE conference 'The Science of Making Torque from Wind', 2012.

All these observations have led to the awareness that wind energy is a technology and application by its own, building on several other fields such as meteorology, aerospace, offshore oil and gas, and power systems. The research challenges described in this long-term R&D agenda refer to this integrated wind energy field of technology, research and application.

The Impact on the Cost of Energy (CoE)

The CoE is a main driver for new developments in technology and design. The definition of the CoE may change over time. At this moment the CoE is defined as the costs of construction, installation, maintenance and finance divided by the energy yield. If, for example, power quality and controllability is accounted for in the CoE, designs will change. Furthermore, the scale of the system under consideration matters (e.g. a single wind power station or the total of offshore wind power), as well as whether life-cycle-based costs are used or only investment and operational costs, and similarly whether the environmental and sustainable values for society are included or not. These choices in defining the CoE will have consequences for technology and design. Long-term research challenges concern pre-competitive research. Its impact on the CoE becomes visible only after implementation of the results in technology and design.

The 11 research challenge descriptions present the questions to be solved in order to make significant steps. These research areas have their own objectives, but all interact with progress in the other disciplines. As the development of wind energy requires a multidisciplinary approach, it is the combination of several areas that makes things happen and that often defines new scientific challenges. In Sect. 7, on design methods, it is discussed how to integrate the knowledge from several disciplines, and how to optimise the design of a wind turbine and power station for the lowest CoE.

An Integrated Approach

Wind power stations have to fulfil the "L^3 conditions": low cost, long lasting, and low service requirement. Design and optimisation for only one of these objectives is not efficient. As an example the effect of turbulence on the overall power output and power quality is mentioned:

- The characteristics of the inflow turbulence need to be measured in real time;
- The conversion of inflow at the rotor to torque and rotational speed of the drivetrain must be known in detail to optimise rotor and drivetrain control, and assess the remaining service life;
- The aerodynamic interaction of every turbine with all others must be quantified;

- The control of rotors, drivetrains, power electronics and power collection in a transformer station needs to be tuned and optimised;
- Finally wind energy must be understood as an integral and active part of a large interacting energy supply system, from a technical as well as economic and societal point of view.

The objective of the optimisation is a maximum yield coupled to power quality requirements of the grid operator, with at the same time, the lowest possible fatigue and extreme loads in the rotor, drivetrain, tower and support structure, and the highest possible reliability of operation. To obtain maximum yield, power curves are used, relating mean wind speed to the produced power. However, the dynamic behaviour of the chain of complex conversion processes is not yet reflected in power curve modelling. Not only for structural optimisation but also for detailed short-time power output analysis, the properties of both conversion processes—from wind to torque by the rotor, and from mechanical to electric power by the generator, converters and grid connection—need to be known in detail.

The Large Scale

The scale of wind energy is increasing, in many aspects. Wind turbines are, on the one hand, the largest rotating machines on earth, with the span of an Airbus A380 equal to the length of a rotor blade, and should be fabricated not as unique exemplars but as mass production (Fig. 3). The size determines the design of all main components, generating new specifications for materials, constructions, control and for the knowledge of wind. It is these new specifications that drive the research questions discussed here. Some examples are as follows:

- Modern offshore wind turbines operate at a more or less unexplored height as it is too high for all other users of the sea (shipping routes, fishery, defence, etc.) and too low for aviation and weather forecast experts.
- As gravity does not scale, the weight of rotor blades and drivetrains (gearboxes, generators) becomes dimensioning, requiring new materials and design solutions.
- Support structures and towers change with scale. Production and installation methods have to adapt to the larger scale and vice versa. Floating turbines offer a solution for deep waters. Floating turbines will enhance the complexity.
- The diverse interactions of one wind turbine in a wind farm with another, and of an entire wind farm with another as well as the entire wind energy production with the earth boundary layer call for new knowledge in turbulence, aerodynamics, control and wind power design.
- Reliability and predictability will become one of the most important design drivers. This has to be considered at a wind farm level, since a wind farm has much more redundancy in sensors, actuators and control than a single turbine.

- Much is known about the effect of a wind farm on man, fish, mammals and birds; much less is known for marine animals. The cumulative effect of several wind parks together on the marine ecosystem is yet unexplored.
- With the increasing contribution to the electricity supply system, the need for international regulatory frameworks increases, as does the need for embedding wind energy in society, taking into account its economic and environmental values.

Fig. 2 The development of wind energy technology during the ages: a very old technology transformed into a modern, state-of-the-art and fully sustainable energy supply system.

Multi-scale Aspects

It is not only the large scale that defines several of the research challenges but also the range from the very small scale to the largest scale that is discussed in the contributions on turbulence, unsteady aerodynamics, materials and machine dynamics. For example, in turbulence and aerodynamics, phenomena at the scale of millimetres (in the boundary layer of the aerofoil) to hundreds of kilometres (determining weather patterns) need to be understood and modelled. The same holds for the nanoscale in materials (e.g. for the bonding of the fibres to the matrix material) to the scale of the rotor diameter, and for the control of aerodynamic devices like flaps of 1 m to the control of the output of the wind power station. For most of the scales models do exist, but these need refinements and tuning for the wind energy specific requirements. Moreover, the combined use of the detailed knowledge on all scales is not feasible and manageable in a consistent way. Rigorous methods to bridge scales are needed in order to distinguish what information is important what can be neglected, and what uncertainties are caused by simplifying approaches.

Big Data and Relevant Information

With state-of-the-art methods, a tremendous amount of information is available to support research and to help the design and operation of a wind power station. Through the use of lidars and buoys, wind and wave data become available, upstream as well as within the turbines. All turbines permanently collect huge numbers of operational data from hundreds of sensors, ranging from gearbox oil temperature to stresses in the blade root. All control actions use sensor data as input, with increasing resolution and complexity. Condition monitoring and assessment of remaining service life rely on clever processing of these data. Power electronic equipment measures and modifies the characteristics of delivered power, and adapts the turbines to disturbances from the grid. Proper filtering and quality control of the data are of the utmost importance as the objective is to provide clear messages of the state of the system and to decrease the uncertainty in design and operation, instead of increasing this.

Validation by Calculations, Model Experiments and Full-Scale Testing

Ideally, the entire wind energy conversion process should be calculated by well-validated computer programs covering the lifetime of the wind turbine or power station. However, this is impossible as it is clear from the following example for the rotor. Applying CFD (computational fluid dynamics) and FEM (finite-element models) to calculate in detail the aerodynamic and aeroelastic behaviour of the rotor for all conditions that the wind turbine will meet during its lifetime is far beyond the capability of these models and will last longer than the turbine's lifetime. The only way forward is to synthesise sample calculations, dedicated experiments and physical knowledge into models for design and analysis. The models and the underlying knowledge have to be validated by computations and experiments. The wind energy community has invested in fast and accurate wind measurement devices, and in test facilities for blades, nacelles, drivetrains, support structures and other components. Many of these facilities perform full-scale testing. Model testing is done in many laboratories, of which wind tunnels are striking. Many of the new wind tunnels in Europe built in the past decade are designed for wind energy. However, the increased size of the wind turbines makes it complicated to reach the right model numbers, most notably the Reynolds and Mach numbers, making it necessary to also develop expensive purpose-specific high-pressure or cryogenic wind tunnels. Component and scale testing and full-scale prototype testing for on- and offshore turbines are available.

Besides all of these achievements, a common strategy is missing where the most relevant and challenging aspects for testing and validation are worked out scientifically and where common blind tests are set up. Such blind tests should concern not only the verification of models and simulations but also the experiments at different locations. An essential part of this is that all results should become finally freely available to the whole scientific wind energy community.

The different approaches of fast, low- and medium-fidelity modelling, expensive high-fidelity calculations, and laboratory and full-scale measurements are complementary. In fact, each of them has benefits and drawbacks related to aspects like the degree of modelling detail, space and time resolution and statistical convergence. The design and execution of a good validation of models is not straightforward. Only the combination of the approaches and analyses of the results can give confidence and reliability to the validation.

Wind Energy Needs Science

Wind energy has become a mature field of technology, with many developments lying ahead. To realise these, dedicated knowledge is required, spanning from basic fundamental science to pioneering technologies, across multiple disciplines. Next to the overview presented above, the research challenges in each of these disciplines will be presented in the following sections.

All sections have the same structure: after a description of the state of the art, several main research questions are formulated, detailed in one or more subtopics. For each main research question a research challenge is formulated.

Fig. 3 Wind turbines are the largest rotating machines on earth.

Chapter 1
Materials and Structures

R. Nijssen and D. Lekou

State of the art

Fibre-reinforced plastics used in rotor blade composites are characterised by the fact that they are exposed to extreme high-cycle fatigue loads, while architecture and manufacturing methods are to a large extent driven by investment cost.

A second main characteristic is that, for these materials, very few dedicated models for predicting fatigue life exist, whereas knowledge about experimental techniques and structural health monitoring is also limited.

Developing optimised accurate and reliable prediction schemes that are also generally applicable has been a long-term research topic. However, the nature of most methods applied today is descriptive or phenomenological, and these methods are not based on deep insight into material intrinsic failure mechanisms. Structural health monitoring has been developed using various technologies and software tools, but the explicit relation between damage state and remaining life is often unclear at present.

Rotor blade manufacture comprises modern materials and production methods, but materials performance improvement is driven by incremental improvement of existing technology for matching and accommodating continuously increasing structure dimensions (Fig. 1.1). The potential of new materials is thereby probably not utilised to the full.

For substructures, a combination of high loads, materials fatigue and cost closely interlinked with the sheer size of support structure components makes these factors the main design drivers. Component connection methods and mass optimisation related to meeting large-scale production and installation demands are a potential (fatigue) failure source.

R. Nijssen (✉)
Knowledge Center WMC, Wieringerwerf, The Netherlands
e-mail: r.p.l.nijssen@wmc.eu

D. Lekou
Center for Renewable Energy Sources, Pikermi, Attiki, Greece

G. van Kuik and J. Peinke (eds.), *Long-term Research Challenges in Wind Energy - A Research Agenda by the European Academy of Wind Energy*, Research Topics in Wind Energy 6,
DOI 10.1007/978-3-319-46919-5_1

Fig. 1.1 The design and manufacturing of rotor blades requires full knowledge of fibre-reinforced load carrying structures

The interface between turbine foundation and soil interface can be regarded part of the support structure, but soil properties as well as their specific degradation patterns over time are still largely unknown. Similar to rotor blades, specific wind energy characteristics do not allow for simple extrapolation of known principles and insight into load patterns from other well-developed applications, and often create new challenges instead.

Finally, materials related and structural issues specific to wind power application include main components and systems together with drivetrains, as well as their lubrication performance, essential for long-term installation upkeep.

1.1 How Will Multi-scale Modelling Improve Materials and Structures?

For any material used in load-carrying structures, there are by definition scale effects involved. These are often addressed through using simulation models applied on different scales for accessing information that would not be available otherwise and for analysing and interpreting experimental results. Especially composite-material-based structures require multi-scale modelling, such as modelling materials at macroscopic level as homogenous and uniform, whereas at smaller scales they could have a distinct non-uniform and/or anisotropic structure.

Lack of a systematic multi-scale modelling approach has already hampered appropriate material characterisation and development of material models for several decades. Improved understanding of the relationship between microstructure and macroscopic material behaviour would provide the means for considerably advancing material technology in the fields of design, manufacturing and validation.

For a multitude of materials, but notably composite materials, their long-term behaviour is not well understood. Current insight in these single and/or combined fatigue and environmental loading effects is largely based upon semi-empirical models rather than a profound understanding of the underlying mechanisms and their interactions.

Scientific challenge 1-1: To explain, through microscopic behaviour, the long-term macroscopic performance of composite materials.

1.1.1 Missing Links in Current Damage and Failure Process Predictions?

For the composite laminated structures of rotor blades, one can make use of multi-scale analysis, firstly for an appropriate understanding of and representation of the failure process, and also for the further development of efficient, engineering methods, in which 3-D effects (such as 3-D stress states and delaminations) are appropriately included. Micro-mechanical, multi-axial, (X)Finite Element Method (X)FEM materials' behaviour assessment should be combined with intensive testing for improved understanding of failure and damage growth processes. It should thereby be noted that for dealing with many damage mechanisms, the existing methods are either working well or promising. In the case of the understanding of long-term mechanisms involving fatigue, current analysis speed is limited to unrepresentatively small-scale or coarse homogenisation only. Thus, dominant or interacting damage mechanisms might not be represented correctly (in the case of too small-scale representation), or not represented at all (in the case of coarse homogenisation). A related question is whether the optimisation methods should be accelerated numerically and/or via hardware tools. Finally, how can such experiments be supported best for the development of numerical simulations, and how can one deal, for instance, with scatter in experimental fatigue on constituent materials?

1.1.2 Multi-scale Experimental Observations?

Various experimental microscopy techniques are available, but most are destructive or invasive methods hampering the monitoring of damage progression on small-length scales. Online high-resolution observation of large volumes of material using, for instance, tomography techniques are under development. Such monitoring is required not only in test laboratories but also during actual composite manufacture.

1.1.3 Linking (Microscopic) Damage State with Macroscopic Observations into Health-Based Strategies

Materials macroscopic state or performance should be measured appropriately. Current health-monitoring techniques are often limited to local observations, while full understanding of the feedback such as the relation to measured and (microscopic) damage state is lacking.

1.2 How Can New Materials Be Developed?

Natural designer response to resolving a specific problem is relying on experience together with utilising materials and manufacturing techniques that are both known and available. Structures typically exhibit discrete material transitions where gradual transitions would be optimal. As long-term reliability and the cost of structures become increasingly critical, the requirements for developing new materials become more pressing too.

Scientific challenge 1-2: To target performance/cost ratio with new materials and manufacturing methods.

1.2.1 Materials Science

In terms of chemistry, the performance of polymer matrix composites used in rotor blades depends largely on the interface between fibres and matrix. A main question is whether there is still a gap in thinking between chemists responsible for interface design and blade designers primarily focused at its performance characteristics. Should the possibility of improving essential material properties by the addition of nanoparticles be investigated? In addition, polymer chemistry is to a large degree determined by state-of-the-art vacuum-infusion processes. An overall question is whether industry is on the right track with the current approaches.

Another issue is that offshore wind turbines operate under harsh environmental conditions whereby protective material coatings and retaining long-term overall quality represents a substantial cost. Could modern chemical processes therefore be utilised for developing new materials capable of diminishing wear, corrosion, bio-fouling, while having to deal continuously with extreme loads and (fatigue) load variations?

In terms of metallurgy, the manufacturing of thick-walled steel sheets like those used for constructing offshore support structures has specific limitations and impact on materials performance. The same applies for bearings, gears, bolts and other fasteners, as well as wind turbine forged shafts and castings such as for the hub and main carrier.

Is the state of the art sufficient, or should new materials and technologies be developed instead, such as bio-resins for plastic composites, wooden towers, or a composite instead of cast iron hub? Known bio-based materials, for instance, require additional, often poorly understood processing steps, and their application includes questions on environmental impact and sustainability performance.

Multi-functional materials include controversial rare earth metals used in permanent magnet generators and multiple additional applications including power electronics. Here, there are questions regarding availability, price development and price volatility risk, and supply chain sustainability.

Utilising such multi-functional materials is a trend but often in parallel represents at least one and often multiple design constraints.

An overall question is whether it is a good idea to use such materials on a substantial and increasing scale.

1.2.2 From Material to Application

Here, a fundamental question regards the validity and correctness of material-related models and whether these are ready to incorporate new materials and/or new material functions.

A second question is on their capability to correctly predict long-term behaviour of structures exposed to multi-axial and varying load conditions, and what this means for new materials. Post-processing and optimisation methods should thereby enable optimal design in terms of performance and cost functionality.

The difficulty in answering these questions is that in going from a material to an application, at a certain point the material is incorporated in a structure. Thus, multi-material interfaces occur, which may be poorly modelled. Finishing operations may influence material behaviour. New stress concentrations may arise. In wind energy, recent answers to these questions are to carry out FEM analyses and to test ministructures or subcomponents. A fundamental challenge is to determine a priori which of these is required and to what extent.

In the long term this is a question that should result in guidelines from a certification body. In some industries, for instance, verification is part of certification rules, giving guidelines to test material on coupon level, scaled or full- scale tests on product level, and subcomponent tests in between. It is further envisaged that, in the long run, the scientific community will succeed in making various advanced predictive tools available to the engineering community. One of these envisaged tools is capable of accurately predicting structural responses, taking into account the response processes from micro-mechanical level within a material to the macro-mechanical level of the structure itself. It considers manufacturing processes affecting every level in a loop, and is accomplished through detailed modelling and reduction of information as the complexity level is being increased. Powerful computing systems as a main process acceleration enabler will allow simulating a 20–30-year structure design life in a reasonable time span through advanced probabilistic methods.

1.3 How Do Joints Really Work?

All structures, whether in composite, metal or concrete, require joining and joints to connect various parts, components and systems. Joining methods are in fact already essential and their importance increases continuously, both physically and at design level. For wind turbine structures, joining components presents a major challenge due to the size of individual elements to be connected and the loads carried by structural parts. Large size not only affects material properties per se but also increases actual tolerances and tolerance fields, leading to multi-axial loading conditions not encountered in smaller size structures. This observation applies to composite-material joints (adhesion, bolts) and joints featuring hybrid connections between two types of materials, currently still rather new to the scientific community, as well as to traditional joints used in civil engineering structures, including welds and grouted and bolted connections. With any joint, a trade-off must be made between a desire for high safety factors, set against unwanted mass and cost increments for mass-produced wind turbine structures.

Scientific challenge 1-3: To better understand and predict the response of a (hybrid) joint under multi-axial fatigue loading.

1.3.1 Connections

Can failures in adhesive joints of composite-material structures exposed to multi-axial fatigue loads be well predicted? Are we thereby limited by bolt standards, and what happens if bolted connections point to bolt sizes that exceed "traditional" dimensions? How can predictions for these sizes be improved? May hybrid materials in load introduction areas improve the structural efficiency?

Are current estimations adequate for the new materials available for large-size grouted connections and can failure be predicted when connections are exposed to both axial and bending fatigue loads?

Finally, what are the primary challenges linked to the welding of very thick-walled steel sheets and other components? (Fig. 1.2).

Fig. 1.2 Support structures: thick-walled welded steel constructions

Chapter 2
Wind and Turbulence

J. Mann and J. Peinke

State of the art

Optimal wind turbine placement requires deep insight into prevailing wind conditions at any relevant position of a terrain or space, from flat land to forested regions, urbanised sites, mountainous areas, or offshore alike. Predicting and characterising wind turbine operating conditions requires detailed knowledge of aerodynamics and atmospheric wind fields down to millimetre scale. Basic equations for atmospheric and aerodynamic aspects are well known, but exact airflow modelling is impossible. This basic problem is due to the interaction of flow features of many different scales and the fact that these equations are nonlinear and non-local. To grasp all different aspects many different approximations have been developed, which can be roughly divided into three different approaches.

The most pragmatic approach is to try to solve the Navier–Stokes equation through numerical methods, called computational fluid dynamics (CFD). For the given scales and Reynolds numbers this is not possible in a direct approach, and thus different approximations have been set up, like mesoscale models for meteorological aspects or Reynolds-averaged Navier–Stokes (RANS) equations for aerodynamic aspects, to mention but the most popular models. These general solutions are known to be chaotic, meaning they are sensitive to the initial and boundary conditions. As smallest errors grow exponentially with time, any approximation inherently becomes a compromise with its assumed simplification. In the investigation of wind flow, interesting patterns such as vorticities, streaks, or gusts can be observed. In particular for transitions regions, where the wind flows evolves after perturbation, it is common to use coherent structures to grasp its

J. Mann (✉)
Technical University Denmark, Kgs. Lyngby, Denmark
e-mail: jmsq@dtu.dk

J. Peinke
Carl von Ossietzky University Oldenburg, Oldenburg, Germany

G. van Kuik and J. Peinke (eds.), *Long-term Research Challenges in Wind Energy - A Research Agenda by the European Academy of Wind Energy*, Research Topics in Wind Energy 6,
DOI 10.1007/978-3-319-46919-5_2

structure. The central challenge will be to understand the generation processes, the lifetimes and the decay mechanisms of these structures.

Statistical methods are commonly used to characterise the high degree of disorder in a turbulent flow. Stochastic processes and/or multifractal scaling properties are used to grasp the multiscale properties as well as to set up models for these flow condition.

The inherent nonlinearity and chaos of fluid dynamics occur at all scales, from weather patterns relevant to wind power grid integration, to turbulence essential to dynamics turbine loads. The inability to calculate flows from first principles has made continued interaction between modelling and measurements indispensable. Although powerful computers now enable simulating turbulent wind conditions with increasing detail, precise results are not guaranteed. As flow simulations are very costly, ongoing additional efforts in large-scale simulations and manageable surrogate wind fields are required for investigating turbine-operating dynamics.

2.1 How Should We Characterise the Dynamical Inflow Conditions?

The characterisation of wind conditions is commonly conducted by using well-defined IEC 61400 standard parameters, all highly relevant for a wide range of different wind power related aspects. These parameters range from average wind speed data for securing project finance to local forces acting on a blade directly related to prevailing local wind inflow. They further compromise different statistics like those of turbulence degree or wind speed distribution, as well as for localised conditions like wind gusts, which are not taken into account by other "normal" statistical parameters. A closer reflection on the choice of different parameters reveals that a rigorous approach to characterise wind conditions is missing. Nevertheless, all these environmental parameters are applied for the wind turbine design (Fig. 2.1).

Scientific challenge 2-1: To work out key parameters across scales that enable a rigorous approach to characterise wind conditions.

2.1.1 Parameters for Wind Turbulence

A challenging question is how to model and characterise wind and turbulence, but it is questionable whether commonly used parameters are the most relevant. Another question is whether there are still "missing" additional environmental effects impacting turbine operation apart from those included in current standards and, equally important, how these are characterised. A new set of parameters is therefore required for specifying wind turbine operating dynamics in greater detail. The main issue of finding relevant key parameters thereby has to be defined from a wind

Fig. 2.1 Wind energy needs meteorological knowledge at altitudes not explored in depth before

turbine perspective and has to take into account meteorological as well as the pure fluid dynamical issues. This in turn requires employing the latest insights about the overall process of converting wind energy into aerodynamic, mechanical and electrical wind power, all at turbine level. One clear example is a particular ambient temperature and wind speed combination with specific impact on a given turbine area. Other environmental conditions like wave swept moisture particles, aerosols impacting hub and blades, and icing conditions have to be taken into account, too.

2.1.2 *Pattern of Wind*

In the above-mentioned standards the characterisation of wind conditions is supplemented by an assumed gust pattern. Investigations of wind field show clearly that there are many flow structures, most prominently extreme events often taken as gusts. Although there are many reports on structures, a commonly accepted definition of a gust is missing. Is there a difference between big gusts and low-level jets? There are also patterns generated by orographic conditions or by wind turbines themselves in their wakes. Other examples could be a particular wind speed or weather pattern with a special effect on the working states of a turbine or on the grid stability. These and other conditions and their individual and cumulative impacts have to be covered by a classification scheme.

2.1.3 Orographic Dependences

The characterisation of inflow conditions for wind turbines has to take into account not only different meteorological conditions but also different environmental conditions. As wind turbines are being installed more and more in complex terrain and offshore, the question is how to generalise an inflow classifications scheme so that many kinds of different locations can be well characterised.

2.2 What Is the Importance of Open Questions on Turbulence?

Within the wind energy context, turbulence means fluctuations in wind flow speed from a scale of kilometres down to a few millimetres and a timescale of hours down to a fraction of a second. For atmospheric turbulence there is an issue of the transition from a quasi two-dimensional large-scale structure to a three dimensional small-scale wind structure, which has led to a debated spectral gap concept. Within this context, wind turbines with sizes of about typically 100 m represent small-scale objects.

The purest form of turbulence is described by the Navier–Stokes equation and considered homogeneous and isotropic. Many flow models are based on the assumption that turbulent wind flow on large scales approaches, at a sufficiently small scale, a universal small-scale turbulent state, which is reduced to the simple Kolmogorov model of 1941. It has been known since at least the 1960s that some turbulence features are not universal at all. Most prominent is the turbulence intermittency problem, which states that the smaller the scales, the more violent turbulent wind fluctuations become. This intermittency problem remains to present a general open question and is not fully taken into account for common flow simulations, which may have important consequences for wind turbine operating conditions.

Scientific challenge 2-2: To understand small scale turbulence for instationary atmospheric and complex orographic conditions.

2.2.1 Small-Scale Turbulence

Small-scale turbulence is still considered to be one prominent unsolved scientific problem, which indicates that details of its structure are not perfectly known. Nevertheless, in the last few decades many details have been worked out under idealised laboratory conditions, which are typically well defined and stationary. Even for this case it is unclear whether or not turbulence is a universal phenomenon. This question is closely linked to the question how to fully characterise a turbulent flow. The often-used turbulence intensity and second-order correlations have to be supplemented by higher-order statistics. Considering the application for wind energy, is there a closure which allows for truncation of the statistical complexity?

Considering real turbulent wind situation, there are serious doubts on whether small-scale turbulence is indeed universal, whereby key details might depend upon environmental variables and details. For complex terrain like mountainous or forest regions it remains unclear which details should be taken into account and what others might be condensed in specific situations. Furthermore, instationarity is expected to play an important role, and thus the question arises of how turbulence changes with time and space.

2.2.2 Structures Within a Turbulence Situation

For the characterisation of turbulence there is a continuing debate on statistics versus coherent structures. It is commonly known that turbulence will create a disorder with anomalous statistics, i.e. with many extreme events. Are gusts just the consequence of small-scale turbulence or do gusts have more features? It is evident that extreme events have more variability than the standard 50-year gust, but what are the most important aspects that could improve the layout of wind turbines? Are only gusts and extreme events the important structures effecting wind turbines or are there other complex structures too? How important are combinations of wind structures. For ideal turbulence it is known that extreme events can cluster, while often the opposite is assumed.

Another question is the stability, dynamics and the lifetime of such important structures in turbulent wind fields. If, for example, using modern lidar techniques a gust is detected in front of a wind turbine, how does such a structure evolve as it approaches the turbine? To what extent is the stability of a tip vortex or wake flow behind a turbine affected by inflow turbulence? What impact does the initial turbulent flow over a hill have on the flow detachment behind this hill? This complex wind flow relation is essential for calculating the wind resource further downstream. In a more general way, one may ask, how does turbulence interact with turbulence? Turbulent wakes might, for instance, interact with atmospheric turbulence, a topic today discussed under the label wake meandering; this describes the impact of another turbine on turbines operating in the wake. This turbulence–wake interaction as a phenomenon has huge importance for wind conditions within a wind farm. A primary issue arises here from the fact that, for such flow situations, the superposition principle does not hold due to intrinsic nonlinearities. A major challenge therefore regards how to add up turbulent effects.

Another challenging question, which we will mention just briefly, is how turbulence produces and propagates noise.

A final question is on the role of atmospheric stability. All these aspects are key for understanding a wind turbine's operating conditions and dynamic loads.

Fig. 2.2 Understanding and modelling turbulence still is a major challenge in fluid dynamic research

2.2.3 Turbulence Validation

It is common to use turbulence models for the design of wind turbines. What is generally required for adequate validation of such models? Model validation is essential due to the fact that no exact solution for highly turbulent operating states is known (Fig. 2.2). Because all measuring systems are limited by spatial and/or temporal resolutions, a question has arisen about what could be most relevant for a measuring campaign. Choosing the incorrect instrumentation and/or wrong location might both result in useless conclusions.

2.3 How Can One Model Wind, as an Energy Resource, in Space and Time?

To get the most benefit from wind energy it is not only necessary to characterise correctly the wind resource conditions but it is also needed to achieve the most precise modelling. Such models will demand a correct description of small scale local wind structures up to climate change. The underlying physical equations will be the same in principle for all scales, but as it is impossible to solve them exactly, thus models are needed that encompass a special ranges of scales.

For large scales it is of central interest to derive predictions. The evolving climate of the next decades will be important for the planning of bigger wind farms in order to be able to estimate the power performance and, thus, the financial risk. Weather forecast is becoming more and more important for grid management and the efficient supply of power. For the operation and control of, wind farms and single wind turbines, the prediction of specific wind situations is also of growing interest. The shorter the scales become, the less important it becomes to correctly predict the momentary properties of special wind situations, and thus it becomes more important to get such issues correctly in a statistical sense. For all these aspects, a big question is how to achieve the best simplifications, which are often done by neglecting less important scales. Another approach is combining different simplified models in the right way. Besides the technical problems such as nesting computational codes designed for different resolutions, a general theoretical basis for such truncations or extensions is missing.

Scientific challenge 2-3: To be able to set up the correct reduced models based on the right scales to grasp all important features of wind fields in space and time.

2.3.1 Forecasting Weather and Climate

Meteorology studies long-term climate effects, as well as seasonal and annual prediction of local wind conditions around a wind turbine. Common challenges include prediction accuracy and necessary detailing, whereby especially the prediction of effects at a range between 0.1 and 1 km remains very difficult. As computer performance is continuously increasing there is an ongoing opportunity as well as remaining challenge to improve weather models.

Multi-scale interaction is a central theme of meteorological modelling. How do large (1000 km) scales thereby interact with turbine site wind flow? A primary and very practical challenge is how to apply mesoscale flow modelling through weather forecast models for wind flow at a particular site, but inherently complicated by all surrounding in-homogenous impacts. This issue concerns both downscaling and nesting, but there is at present no consensus on a preferred and scientifically most correct procedure. It is unclear whether it is necessary to also incorporate effects of backwards cascades, in which small-scale structures influence larger scales.

The following issues exemplify the need for an improved understanding of multi-scale interaction:

- the extraction of momentum (drag) of a complex terrain represented in a meteorological model;
- the understanding of extreme meteorological events obtained by mesoscale models, whereby the most challenging extreme events to be predicted are not the largest but rather those of medium size occurring at scales up to a few kilometres, including, in particular, local events like thunderstorms and downbursts.

Traditionally, mesoscale models are evaluated on their ability to predict temperature, cloud cover, precipitation and other parameters of little relevance for wind power generation. There is still great work to be done for improving on capabilities of these models for specific wind energy use.

2.3.2 *Limits of Predictability*

The quality of any forecast is based on an estimation of the intrinsic errors, which are based on the nonlinearity of the problem itself. A second aspect is to what extent the assimilation of wind measurements, such as from wind turbines, could improve forecasting. A third question regards whether tipping points can be found for specifying the overall quality of predictability. How can the precision of the forecast of turbulence be quantified?

From a physical point of view it is thus essential to realise that an inflow model conserves vorticity, facilitates energy transport, and/or is divergence-free in a phenomenon also called solenoidal vector field. How important are these constraints for the quality of predictability.

Chapter 3
Aerodynamics

J.N. Sørensen and C. Ferreira

State of the art

Wind turbine aerodynamics is a central discipline for modelling and prediction of the aerodynamic forces on a wind turbine. From the aerodynamic analysis the performance and loads on the rotor blades, as well as other structures exposed to the wind, are determined. An aerodynamic model is normally integrated with models for wind conditions and structural dynamics. Integrated aeroelastic models for predicting performance and structural deflections are a prerequisite for the design, development and optimisation of wind turbines. Aerodynamic modelling also concerns the design of specific components, such as rotor blade geometry, and systems of wind turbines, such as performance and optimisation of wind farms.

Wind turbines are subjected to atmospheric turbulence, wind shear from ground effects, wind directional changes both in time and in space, and wake effects from neighbouring wind turbines. Vertical-axis wind turbines also experience blade–vortex interaction and streamlines which cross the actuator surface twice. The combination of these effects represents the normal operating conditions experienced by the blades. As a consequence, the forces vary in time and space and a dynamic description is an intrinsic part of an aerodynamic analysis.

A challenge for analysing wind turbines operating in the atmospheric boundary layer is the very different length and timescales involved. These range from the millimetre-thin boundary layer on the blade surface to the length of modern blades, which are now approaching 90–100 m. Considering wake effects and the mutual interaction between wind turbines and wind farms, the length scales range from a few to hundreds of kilometres. This is illustrated in Table 3.1, which shows the various scales related to wind turbine aerodynamics.

J.N. Sørensen (✉)
Technical University Denmark, Kgs. Lyngby, Denmark
e-mail: jnso@dtu.dk

C. Ferreira
Delft University of Technology, Delft, The Netherlands

G. van Kuik and J. Peinke (eds.), *Long-term Research Challenges in Wind Energy - A Research Agenda by the European Academy of Wind Energy*, Research Topics in Wind Energy 6,
DOI 10.1007/978-3-319-46919-5_3

Table 3.1 Scale requirements in wind turbine aerodynamics

	Length scale (m)	Velocity scale (m s^{-1})	Time scale (s)
Aerofoil boundary layer	0.001	100	0.00001
Aerofoil	1	100	0.01
Rotor	100	10	10
Cluster	1000	10	100
Wind farm	10,000	10	1000
Cluster of wind farms	100,000	10	10,000

Currently, it is impossible to deal with all these length scales in one computation; substantial research has been conducted in developing techniques that allow combining different scales by sacrificing the details of the flow. An example is steady RANS modelling, which allows calculating cumulative wind turbine flow comprising the individual flow around the blades and main components such as the tower and nacelle. Physical phenomena that still remain uncaptured are boundary-layer details, such as the laminar–turbulent transition, which is typically predicted through semi-empirical relations. Another example is the actuator line model, which enables conducting large eddy simulations of wakes without resolving blade aerodynamics in detail.

The spatial and timescales also limit the feasibility of experimental research. Field tests are limited due to cost and complexity of the setup and operational conditions, while experiments in controlled conditions (e.g. wind tunnels) usually fail to replicate the scales of the flow (e.g. Reynolds number, reduced frequency).

In parallel to the numerical challenge, requirements for experimental research are currently beyond the capability of most experimental facilities. In addition, the requirement of increasingly larger facilities is often not feasible or too expensive. A main limiting parameter is the Reynolds number, which has now reached 10^7, while most wind tunnels are restricted to operation at Reynolds numbers of up to 5×10^6. It is noteworthy that the same constraint also exists for numerical simulations, as the required number of grid points approximately scales as the square of the Reynolds number.

3.1 Is the Acceleration of CFD Codes the Main Challenge, or Do We Still Have Physical Problems to Solve in Rotor Analysis?

The continuous growth in computational power is being matched by increasing demands for complex aerodynamic phenomena to be simulated.

CFD and wind tunnel experiments are powerful tools for rotor analysis in design and off-design conditions, but we are still far away from the possibility of understanding all spatial–temporal details, which in principle may be of importance for the aerodynamic situation. Apart from turbulence modelling, severe off-design conditions and very strong fluid–structure–control interactions, calculations and experiments agree reasonably well. A question here is on whether the reduction of CFD process time and higher resolving experimental methods are the single remaining problem open for research. Which are the essential processes in terms of scales and physical models for the energy conversion process? Which models are suitable for aerodynamics with body interaction? How can one overcome the challenge of scales in the (experimental) validation of models?

Scientific challenge 3-1: To work out rigorous methods in order to be able to exploit the growing computational power and experimental advances for clarifying challenging problems in aerodynamics.

3.1.1 Improvement of Simplified and Low-Fidelity Models

Several of the most effective aerodynamic design and analysis tools combine several high- and low-fidelity models. In parallel to the reduction in cost of CFD, improving the accuracy of simplified models is one of the key scientific and engineering challenges for wind energy.

Many engineering rules used in the BEM (blade element momentum) model have been verified by CFD and experiments, and the validity of simplified models continues to be assessed. Vortex lifting-line and panel models are increasingly being developed and validated for wind turbine applications, and represent a valuable complement to the BEM model for more complex flows (e.g. unsteady flows, yawed flow, non-planar rotors).

The development of wind turbines and wind farms increases the relevance of unsteady flow aerodynamics, including dynamic stall, due to passive effects (e.g. yawed flow, gusts) and active effects (e.g. flow control, new rotor concepts). Dynamic inflow models (e.g. Pitt–Peters model), unsteady aerofoil aerodynamics (e.g. Theodorsen or Beddoes–Leishmann models), and other models are increasingly used, often beyond their validity. Will these models suffice? What physics are we missing?

3.1.2 Hybrid Models and Eulerian-Lagrangian Formulation

Wind turbine aerodynamics results in inhomogeneous flows. This allows for the opportunity to use simplified models and/or discretisation in certain regions and/or instants of the flow simulation. Hybrid models are currently used in wind energy aerodynamics, usually combining a model for the representation of the blades and near-blade flows with a model for the representation of the remaining flow field, with particular emphasis on the wake (e.g. actuator line models or BEM models coupled with aerofoil aerodynamics models).

Lift-driven wind turbines generate a flow that is dominated by the circulation system, with the circulation on the rotor being convected into the wake. Lagrangian models with discrete vorticity elements (i.e. vortex models) have proven to be effective models for the design and analysis of wind turbines in 3-D unsteady flows. However, these models are limited or too costly when trying to simulate the generation of loads and vorticity over the rotor blade, or when including the diffusion of the wake and the interaction between multiple wakes and with the atmospheric boundary layer.

New modelling approaches are based on hybrid models where, for example, Lagrangian and Eulerian flow field formulations have been merged. However, the transfer of flow properties between the two formulations is still in need of some basic research, in particular with respect to the Lagrangian formulation of turbulence and diffusion. This raises questions on the requirements for a correct simulation in a hybrid formulation. How can, for instance, turbulence and diffusion be correctly modelled in a Lagrangian formulation? And in what manner could multiple scales of a flow be decomposed in multiple models/refinements?

3.1.3 Uncertainty Quantification

The uncertainty in flow conditions, rotor operation, geometry and properties indicates that a deterministic simulation is insufficient to support design. Methods that apply a statistical analysis of uncertainty through models have been developed in recent years. However, these methods do not provide insight into simulation options and relevant phenomena. What are the relevant levels of uncertainty of different aerodynamic phenomena (e.g. turbulence intensity, circulation)? How can we experimentally and numerically replicate uncertainty? For aerodynamics it is one of the big drawbacks that high-precision results are missing. Typically errors in the range of few per cent are already very stratifying.

3.1.4 Experimental Simulation and Model Validation

Model validation remains a key challenge. Current models are validated either by (1) experimental data in controlled conditions that do not cover the full range of physical phenomena (Fig. 3.1) or by (2) field data for which conditions are not fully described, and for which generality is not asserted. Many models are validated by integral parameters that hide error cancellation. The development of new experimental techniques has increased the amount of measurement points and data acquired. However, we still lack insight into the main phenomena and validation criteria for model validation for all scales, from transition in the aerofoil boundary

Fig. 3.1 Advanced aerodynamic measurements of a rotor blade profile. PIV measurements in the wind tunnel with a turbulence generating grid

layer to inter-wind farm/cluster aerodynamics. What are the key validation parameters for the different phenomena at the different scales? How can one extrapolate from experiments in controlled conditions to full operational range and between scales? How can field data be used for model validation beyond average integral values?

3.1.5 Unsteady Fluid-Structure-Control Interaction

As control and aeroelastic effects become more relevant, unsteady fluid–structure–control interaction becomes an increasing challenge. The development of new concepts (e.g. airborne wind power, flow and load control systems) increases the requirement for accuracy and detail. Structural, aerodynamic and control models usually require different spatial and time resolutions, and the different level of fidelity of the sub-models further impairs the coupling. This raises a number of questions. What are the most relevant scales of interaction, in particular for new concepts (e.g. flow control)? What are the limits of current simplified models (e.g. BEM, lifting line, actuator line) for accurate aerodynamic simulation in fluid-structure-control interaction and how can they be improved? How can one effectively couple models with different fidelities?

3.2 How Can the Aerofoil Concept Be Extended Towards an Unsteady Three-Dimensional Flow?

The aerodynamics of a rotor blade are in general governed by highly dynamic and three-dimensional inflow conditions, which has a substantial impact on the boundary-layer behaviour of the blade. However, for a lift-driven rotor the blade is, for simplicity, typically divided into independent 2-D aerofoil sections, neglecting the influence of the spanwise gradient. In many cases, this approximation can be justified, but near the root or tip of the blade, or for rotors operating near or beyond stall, severe 3-D effects are present. How can one achieve, in 3-D unsteady flow, the necessary accuracy in aerodynamic properties already proven for 2-D aerofoil design, including wind tunnel testing? What is a quasi-3-D or locally 3-D flow? Can the 3-D flow be treated as an uncertainty? Can we improve integral boundary-layer models up to 3-D?

Scientific challenge 3-2: To achieve an understanding of 3D flow pattern and their dynamics on rotor blades including emerging turbulence.

3.2.1 *Flow Separation and 3D Stall*

Three-dimensional effects are unavoidable for wind turbine rotor blades, and are known to play a role with increasing importance for bigger turbines. Three-dimensional effects are highly dynamical, as they are typically generated by instationary inflow conditions. Aerodynamics at large angles of attack, characterised by highly unsteady separated flow, still challenge current models. A 3-D approach is likely required for a correct model, but what are the main 3-D effects relevant for correctly capturing separation, and would a non-stochastic approach perhaps be more effective? What are the main causes for the mismatch between experimental results when viscous effects—in particular separation—become dominant, and how could they be mitigated?

3.2.2 *Roughness, Transition and Turbulence*

Wind turbines operate in an environment that can lead to aerofoil surface degradation. This added roughness can change the interaction of the flow with the aerofoil surface, leading to changes in friction and 3-D surface loading, which furthermore may lead to the generation of 3-D vortices in the surface and higher turbulence. With regard to the impact of roughness and turbulence in transition, an open question is how to model and define contamination-related uncertainty in blade aerodynamics, in particular its effects on transition and separation.

3.2.3 *Flow Control Devices and 3-D Unsteady Flow*

Interlinked with unsteady and 3-D boundary layers are boundary-layer control, both passive (fences, vortex generators) and active (plasma, suction/blowing); dynamic stall; 3-D root region; morphing and control surfaces; closure relations; and transition effects including unsteady flows. All these different fields are addressed by either detailed experimental and numerical simulation or empirical approaches and engineering methods for which validity limits are not yet fully defined. How can the gap between obtaining data and fostering insight be bridged, and what is thereby missing?

3.2.4 *Aerofoil Aeroacoustics*

Aeroacoustics remains an ill-explored field in wind turbine rotor aerodynamics. One of the main sources of noise for wind turbines is located at the trailing edge of the aerofoil, as this interacts with the turbulent eddies of the boundary layer. That aerofoil Reynolds numbers are increasing implies that the experimental and numerical simulation of aerofoil aeroacoustics will become more expensive. Despite several engineering guidelines and empirical formulations, theoretical, numerical and experimental tools available for analysis and design are still limited. The development of accurate computational aeroacoustics models requires additional developments to those in computational fluid dynamics models. A question is whether the current theoretical and experimental framework status is sufficient to support models validation and fully enable aerodynamic design. How can one experimentally measure (the low-frequency) noise in a wind tunnel, determining intensity and location? How can noise reduction devices and surfaces be simulated? (Fig. 3.2).

3.3 How Do Wake and Wake-Wake Interaction Effects Depend on Near-Wake and Blade Flow Details?

In a given wind farm, wind turbines behave as vortex generators that at the same time extract energy and thereby increase the mixing of the atmospheric boundary layer. That in turn increases energy transfer between regions with different stagnation enthalpy. The ability of a wind farm to extract maximum energy yield is not only a function of a given turbine acting as an actuator but also of the vortical structures it generates. Associated with these vortical structures are concepts such as convection, diffusion and stability. The effective modelling of the mixing process in the wind farms might therefore be dependent on the accurate (up to a limit) modelling of these concepts.

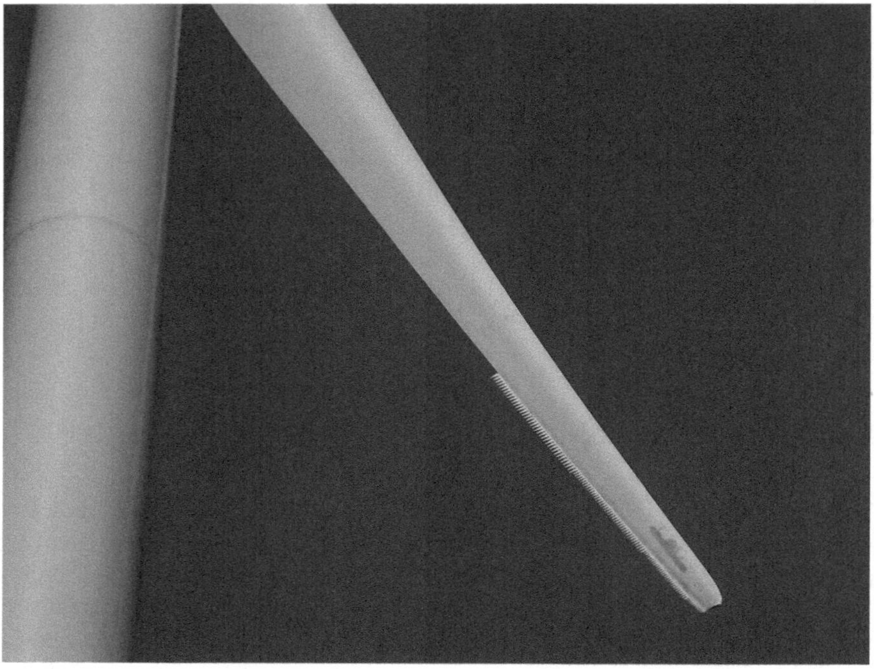

Fig. 3.2 Serrated trailing edge to reduce high frequency noise emission by wind turbine rotors

Scientific challenge 3-3: To achieve a fundamental understanding of the aerodynamics of wakes and interacting wakes behind wind turbines.

3.3.1 Interaction of the Wind Turbine Wake with the Atmospheric Boundary Layer

The stability of wakes is affected by different inflow (turbulence) conditions, asymmetries in tip vortices, rotor off-design situations, etc. What are the relevant scales that must be modelled? How can the effect of the wake on an unperturbed turbulence field be modelled? What types of instabilities may potentially improve the mixing process? A final question is whether it is possible to define regions that are inherently stable and inherently unstable, and, if so, how could this be achieved? See also Chap. 2 and the general discussion therein of structures within turbulent flows.

3.3.2 *Near- to Far-Wake Transition*

Wakes and wake–wake interaction models have a long tradition, based on more or less azimuthally averaged modelling. The transition between the near and the far wake determines not only a change of the structure of the wake but also an increase in the wake mixing process. What are the limitations of current near wake models and how can they be improved, including the effect of turbulence and perturbations? How can near- and far-wake models be coupled? What are the criteria for near-wake stability and how can it be controlled and exploited?

3.3.3 *Wake-Wake Interaction*

Perturbations and instability in the wake are not only generated by turbulence inherent to the unperturbed flow; they can also result from perturbations in rotor loading. This in turn could be a consequence of, for instance, blade–vortex interaction or fluid–structure interaction in partial wake operation or in yawed flow conditions.

Additionally, vortical structures can be considered in increasing clusters of interaction, from an individual vortical structure and/or actuator wake up to the interaction of multiple wakes. This raises a question regarding what information about vortical structures must be preserved for ensuring the correct modelling of their interaction with the actuator. Could multiple interactions simply be treated as turbulence, or are there more dominant stochastic modes and scales to consider?

3.4 Do New and Adapted Aerodynamic Concepts Require New Knowledge?

An overall scientific and wind industrial challenge is that rotor blades are getting bigger, both as a function of turbine size and power rating, as well as in a structural lowering of turbine-specific power rating. Another trend is towards increasingly flexible structures.

The research topics covered are even more relevant for completely new or adapted turbine concepts. Examples include vertical-axis wind turbines, horizontal-axis wind turbines with downwind rotors and/or highly flexible rotors, rotors on floating structures, and airborne wind power systems. For these concepts, the actuator surface becomes three-dimensional and the loads become unsteady. A major question is whether or not current knowledge still applies to these concepts, and the main aerodynamic challenges to be encountered. What are the new aerodynamic challenges posed by these concepts?

Scientific challenge 3-4: To be at the forefront with the required knowledge to deal with new aerodynamic concepts.

3.4.1 Unsteady 3D Actuator Surface

Vertical-axis wind turbines, floating horizontal-axis wind turbines, coned horizontal-axis wind turbines, and (several) airborne wind power concepts (e.g. kites) extract energy by creating a 3-D actuation surface. These non-planar curved surfaces cannot be accurately modelled by conventional BEM models. Could momentum models be extended for 3-D actuator surfaces? How can the unsteady effects and finite blade effects be included? What are the limits of energy conversion for these concepts?

3.4.2 Blade-Vortex Interaction

For 3-D actuation surfaces, blade–vortex interaction can become an intrinsic phenomenon. Are current blade–vortex interaction models sufficient for wind energy applications? How can blade–vortex interaction be included in a momentum model? How can one improve 3-D diffusion models for Lagrangian vortex models? What is the impact of blade–vortex interaction in the stability of the near wake?

Chapter 4
Control

J.W. van Wingerden, D. Schlipf and P. Gebraad

State of the art

The industry's primary goal for state-of-the-art wind turbine controllers is to maximise the energy capture while keeping the turbine within its operational limits, such as maximum rotor speed or maximum power, which is limited by the generator capacity. To guarantee a reliable operation over the turbine's lifetime, the structural loads need to be kept within their design limits. In the turbine and controller design process, an increase in energy is evaluated against higher loads; the result is that designers usually make specifications that increase material costs.

Traditionally, the closed-loop controller consists of slightly coupled single-input–single-output (SISO) controllers, connecting one sensor with one actuator. The main measurements are the yaw misalignment and the generator speed. The yaw misalignment is used by the yaw controller to adjust the yaw position in order to align the turbine with the main wind direction and thereby to maximise the energy capture of the individual wind turbine. The generator speed is used by the torque controller to adjust the rotational speed to its optimal value before the rated power is reached. The generator speed is also the input to the collective pitch controller, which regulates the rotational speed to its maximum value if rated power is reached.

Furthermore, modern wind turbines often have additional SISO loops to reduce structural loads on individual components. These loops include the following:

J.W. van Wingerden (✉)
Delft University of Technology, Delft, The Netherlands
e-mail: j.w.vanwingerden@tudelft.nl

D. Schlipf
University of Stuttgart, Stuttgart, Germany

P. Gebraad
National Renewable Energy Laboratory, Golden, USA

G. van Kuik and J. Peinke (eds.), *Long-term Research Challenges in Wind Energy - A Research Agenda by the European Academy of Wind Energy*, Research Topics in Wind Energy 6,
DOI 10.1007/978-3-319-46919-5_4

- Tower vibration damping: the tower-top accelerations can be used to generate an additional zero-mean collective blade pitch command to damp the tower oscillation.
- Drivetrain damper: an additional zero-mean generator torque command based on the generator speed signal is used to damp vibrations in the torsional drivetrain motion.
- Individual pitch control (IPC): based on measurements of the blade-root bending moments, a zero-mean pitch demand for each blade is added to the collective pitch angle to minimise the pitch and yaw moment of the rotor.

In the state of the art as described above, each turbine in the wind plant optimises its own performance without taking into account its effect on the other turbines. In wind-plant-level control systems, wind turbines communicate with each other to optimise the overall wind plant production and load regulation, thereby assessing interaction effects between the turbines. An important interaction effect consists of the influence turbines have on each other's loads and production through their wakes.

4.1 To What Extent Can Modern Control Theory and Technology Tame the Wind?

For the next generation of (possibly floating) wind turbines, with even larger rotor sizes, a much more detailed and faster adaption to the loads will be required to alleviate the loads on the blades and thus guarantee a lifetime of at least 20 years. Wind is not only the energy source for wind turbines but also the main disturbance to the system. Thus, control of wind turbines is mainly focused on reacting to the changes in the wind flow. New challenges for wind turbine control rise from the increasing complexity of the external inputs; because of larger rotors it is necessary to react to the distributed nature of turbulence in a more appropriate way. Also, the additional impact of waves and currents and the higher flexibility of floating wind turbines need to be addressed. New possibilities arise from information that sensors provide about the disturbances.

Scientific challenge 4-1: To develop and integrate sensors, data processing, control algorithms and actuators, able to handle competing control goals, such as reduction of damage equivalent loads and maximisation of production via *dedicated cost functions.*

4.1.1 Novel Sensor Technologies

Traditional feedback controllers for wind turbines are only able to react when the wind already affects the turbine dynamics. With advances in measurement technologies such as lidar, information about incoming disturbances becomes available ahead of time and can be used for control; however, the disturbance cannot be measured perfectly. This situation requires control research to address two coupled aspects. First, complex information on the wind field needs to be reduced to wind characteristics such as speed, direction, or shears, and a controller is to be developed that addresses changes in the disturbances. Second, the performance of the preview controller depends on how well the disturbances correlate with the actual disturbances acting on the turbine. A thorough understanding of the nature of the wind, as well as signal processing and wind measurement principles, is needed for developing accurate estimation techniques that enable successful preview control algorithms. This requires developing and improving dedicated distributed wind and load sensors with the aim to close the gap between sensor developers and control engineers, and to determine what is possible to measure and which measurements are beneficial for control. Dedicated and robust signal-processing algorithms to provide useful signals to the control system need to be developed; these algorithms will need to take into account changing measurement quality and environmental conditions and need to become integrated in the control and fault detection algorithms.

4.1.2 Distributed Actuation

Active aerodynamic devices and novel sensors, with embedded intelligence distributed along the span, have shown the greatest potential in achieving load reductions and are a proven concept. The development of this kind of technology, often referred to as "smart" rotors, is an interdisciplinary development par excellence. The concept is widely used in airplanes; however, there are still specific challenges that have to be addressed to increase the readiness of the technology for use in wind energy (Fig. 4.1). Reliable actuators and sensors need to be developed and integrated into an efficient controller synthesis method. As the reliability of the rotor may not be jeopardised, algorithms for fault detection and isolation are required. A combination of active and passive control mechanisms (e.g. a blade that twists under bending) may be advantageous.

4.1.3 Floating Structures

In addition to the wind impact, floating wind turbines are excited by stochastic influences from waves and currents and have new dynamics compared to ground-fixed ones (Fig. 4.2). Modern wind turbines use pitch control to limit the power output of the turbine once the nominal power output has been reached. This approach involves a decrease in the thrust force, which, if applied on a floating platform, may cause negative damping of the platform pitch motion. Integration of

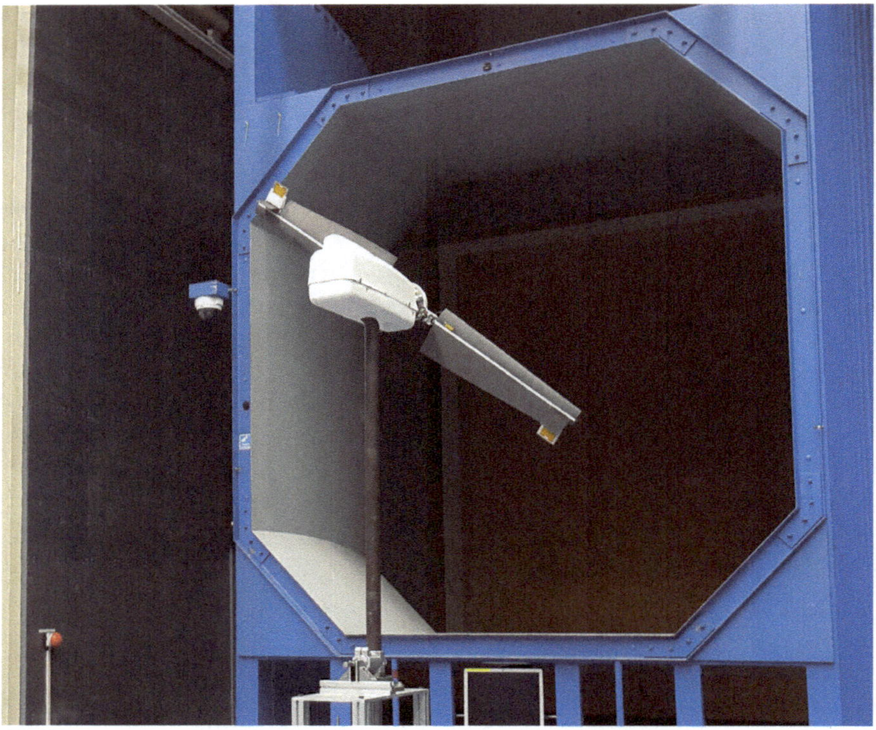

Fig. 4.1 A model rotor in an open jet wind tunnel to test trailing edge flaps for load control

the design of the controller, floating platform and wind turbine will result in a better control performance compared to the sequential design of controller and structure. Sensors (inclinometer, wind and wave measurement systems) and actuators (active and passive dampers) have to be robustly integrated into the control system.

4.1.4 Controller Synthesis

Wind turbine controllers are designed using control-oriented models, and are based on a trade-off between load and power control. Currently, the control algorithm design methodologies used in the wind industry are mainly based on SISO gain-scheduled proportional–integral–derivative (PID) regulators. However, several academic articles discuss optimal control, fuzzy logic control, model predictive control, robust control, and, more recently, linear parameter-varying (LPV) control of wind turbines, and also synthesising self-learning and adaptive controllers, as well as developing data-driven controller design.

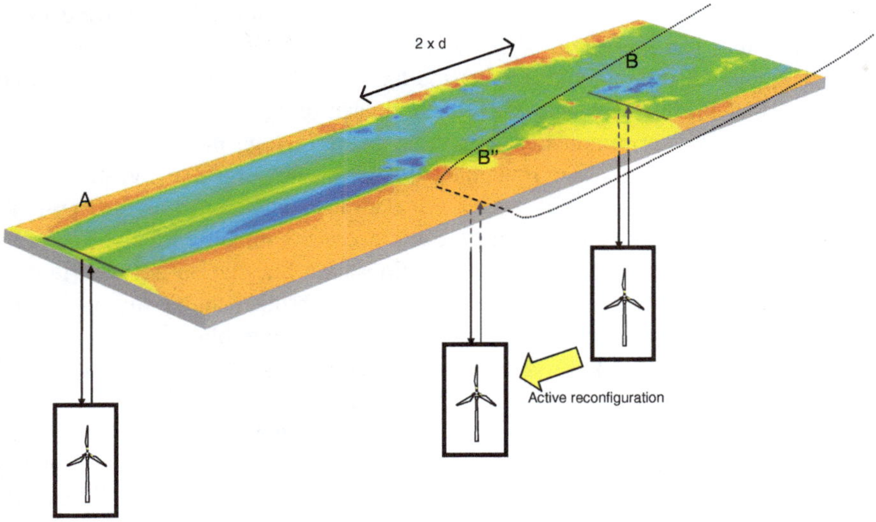

Fig. 4.2 Controlled relocation of floating wind turbine B to avoid the wake of the upstream wind turbine A.

4.2 How Should We Operate a Wind Plant?

Because of the fast growth of the number of large-scale wind plants worldwide, the economic feasibility of the wind energy industry greatly depends on the reliability and maintainability of new wind plants. In a wind plant, the individual control optimisation of each turbine is now mainly done temporally, taking the local (temporal) changes of the wind into account. The optimisation on the plant-scale is of interest when the turbines are integrated in a network, and when they are placed close together such that they aerodynamically interact with each other.

Scientific challenge 4-2: To integrate all relevant aspects, ranging from time-varying weather models to distributed sensing in the wind plant in a control-oriented wind plant model.

4.2.1 Control-Oriented Modelling

In recent years, a large number of wind plant models have been developed. These models have different levels of complexity and fidelity, and may have different purposes (e.g. wind plant performance evaluation, layout optimisation, or controls development).

In the context of wake control in wind plants, a control-oriented wind plant model describes the interaction of the wind turbine with the atmospheric boundary layer. This includes the interaction that the turbines have with each other through the wakes that form in the atmosphere behind each turbine.

At one end of the complexity spectrum, there are high-fidelity wake models based on computational fluid dynamics that are suited for the verification of methods but that are not directly suited for optimisation because of their computational cost. At the other end, there are more computationally efficient control-oriented models that are suited for quick optimisation but that do not describe the physics of the wind plant in detail. Control-oriented, in this context, means that the effects of the wind turbine control degrees of freedom (e.g. blade pitch, generator torque, and rotor yaw) on the wake interaction can be predicted by the model to an extent that is relevant for wind plant performance.

Because wake interaction effects are influenced by the atmospheric conditions (e.g. atmospheric stability and turbulence intensity), the parameters of control-oriented models may be dependent on the specific site and weather conditions. To adjust these time-varying parameters, one can make use of the various measurements in the wind plant (loads and production of wind turbines, and meteorological mast and lidar measurements). The properties of the wind plant itself may be assessed by system identification techniques, based on distributed sensing in the wind plant.

4.2.2 Controller Synthesis

Wake control has mainly been investigated in a centralised framework based on simplifying assumptions, where the effect of a limited number of control degrees of freedom of the wind turbine (e.g. either control of the axial induction of the rotor through pitch or rotor speed control, or wake steering through yaw control) on a limited number of performance indicators (e.g. energy production only) is considered.

Although this current work has shown that significant increases in performance are possible, there is interest in future research into fault-tolerant optimisation of wind plant controls in a distributed framework, which enables scalability as well as improved maintainability and reliability. As mentioned earlier, computational efficiency is an important issue when more high-fidelity models are used as an internal model for the controller. Also, the computational complexity increases as more aspects of wind plant performance are considered in the design. This requires directly synthesising distributed controllers based on high-fidelity wind plant aerodynamic models, with the combined objective of maximising wind plant production and minimising the fatigue loads on the individual wind turbines, in a computationally efficient manner, and incorporating weather prediction and distributed sensing in the wind plant into the distributed controller. A next step is to integrate various aspects of wind plant performance in plant-level control design that have not been considered before, including prediction and measurement of wind turbine fatigue loads, operation and maintenance aspects, grid requirements, and the combined optimisation of wind plant design and controls.

4.2.3 Experimental Validation

Although several large offshore wind plants are currently operating in Europe, most of the academic results in the field of wind plant controls have not yet been validated on a full-scale wind plant. Therefore it is important to develop a statistical framework to evaluate the performance of wind plant controls strategies in a rigorous manner, based on wind plant measurements.

Chapter 5
Electromechanical Conversion

H. Polinder and A. Abrahamsen

State of the art

Within this document, electromechanical conversion refers to the components and systems in between the hub and electrical connections. In state-of-the-art wind turbines, electromechanical conversion mostly consists of a combination of a gearbox, a generator and a power electronic converter. However, different electromechanical conversion systems are being used in different wind turbines as indicated in Table 5.1. Roughly 75 % of the wind turbines have a gearbox; the other 25 % have a direct-drive generator system. Practically all modern wind turbines use power electronics to enable variable speed.

Furthermore, different companies, research institutes and universities are investigating alternative electromechanical conversion systems, including the following:

- hydraulic continuously variable transmission and a fixed-speed directly grid-connected synchronous generator;
- magnetic pseudo-direct drive generators comprising a magnetic gearbox with integrated generator;
- new direct-drive generator configurations, for example featuring a supersize diameter and/or air cored generators or superconducting generators.

During the last decade turbine technologies directly targeting offshore application have emerged with large rotors presently up to 171 m and power ratings up to 8 MW, whereas concepts up to 12 MW and 200 m rotor diameter are in development. These next-generation turbines come with again different drivetrain

H. Polinder (✉)
Delft University of Technology, Delft, The Netherlands
e-mail: h.polinder@tudelft.nl

A. Abrahamsen
Technical University Denmark, Kgs. Lyngby, Denmark

G. van Kuik and J. Peinke (eds.), *Long-term Research Challenges in Wind Energy - A Research Agenda by the European Academy of Wind Energy*, Research Topics in Wind Energy 6,
DOI 10.1007/978-3-319-46919-5_5

Table 5.1 Components of the electromechanical conversion system in state-of-the-art wind turbines

Mechanical conversion	Electromechanical conversion	Electrical conversion
• direct drive (no gearbox) • with gearbox (single-stage, two stage, three-stage)	• squirrel cage induction generator • doubly-fed induction generator • synchronous generator with electrical excitation • permanent magnet synchronous generator	• fully rated voltage source inverter • partly rated voltage source inverter • direct grid connection (no power electronics)

solutions, and must simultaneously comply with more stringent reliability requirements since the maintenance requirement schemes could be different.

5.1 What Are the Physical Limitations and Is It Possible to Shift or Avoid Them?

Understanding the physical limitations of the different components in the electromechanical conversion system may reveal whether it is possible to shift these limitations in order to make these components smaller and cheaper. If it is difficult to shift these limitations, they may be circumvented by using different designs or structures. Alternatively, these limitations may be avoided by using more disruptive technologies that have their own limitations.

For some components of the electromechanical conversion system (such as the gearbox), the physical limitations are reasonably well understood, while remaining within these physical limitations is difficult because the understanding of the system dynamics and the local loads is insufficient. Therefore, the main research focus may be on understanding the system dynamics and the local loads in order to stay within these physical limitations, as will be worked out in the research question "How can the multi-scale problem of the system dynamics and local loads be evaluated?"

Scientific challenge 5-1: To use the understanding of the physical limitations to shift or avoid these limitations and to enable assessment of the potential of the different possible drivetrain components.

5.1.1 Mechanical Conversion Systems—Gearboxes

The physical limitations of a gearbox are mainly related to the stresses on the gear teeth and the bearings caused by the local loads. For shifting these physical limitations, better materials that can deal with higher stresses would be necessary. These materials are not readily available. A long-term research question is whether it is possible to develop these stronger materials.

It is difficult to stay within these physical limitations, as appears from the fact that especially the gearbox bearings regularly fail. Like for other components of the electromechanical drivetrain, this is because of unknown (local) loads, coming from the whole-system dynamics (including the wind, the electric system and the control). This is aggravated by the fact that teeth and bearings may not be uniformly loaded because of manufacturing tolerances of the gears and the bearings and because of the deformations and displacements of the drivetrain components due to high loads. When scaling up, it becomes increasingly difficult to stay within the physical limitations.

The local loads and stresses are highly dependent on the gearbox design. Therefore, different gearbox structures are developed to increase the torque levels without increasing the local loads and stresses. Because of the high input torque, almost all gearbox manufactures are using gearbox structures with load-sharing functionalities, such as splitting the torque into parallel paths and increasing the number of planets in a gear.

Nevertheless, some wind turbine concepts are leaving out the gearbox in an attempt to face the mentioned problems in another way. The following are example of these disruptive technologies:

- Hydraulic torque conversion systems that replace mechanical gearboxes. Hydraulic torque converters are promising because they can lead to a reduction in size and weight because the forces are distributed over much larger surface areas than in gearboxes. Systems based on oil as well as systems based on water are being investigated. For oil-based systems, the main challenges are efficiency and reliability. For water-based systems, there is an additional issue with lubrication.
- Magnetic gearboxes that replace mechanical gearboxes. Especially if the magnetic gearbox is integrated with the generator (as is the case in the so-called magnetic pseudo-direct drive), this system can lead to a reduction in size and weight. Two main issues include the use of a large amount of permanent magnetic material and the complicated construction with an inter-rotor.

5.1.2 Electromechanical Conversion—Generators

The power of an electrical machine is proportional to the rotational speed, the shear stress, and the volume enclosed by the air gap. Increasing the speed of the generator makes it possible to convert the same amount of power with a much smaller generator, and that is why gearboxes are used in most wind turbines. However, an increasing number of manufacturers is considering direct-drive generator systems or gearboxes with a reduced number of gear stages.

The shear stress in the prevailing generators is physically limited because it is the product of the current loading and the magnetic loading, which are both limited. The current loading is limited because the heat dissipation is limited, and the magnetic loading is limited because magnetic materials saturate.

During the past century, a lot of research has been done to shift these limitations by investigating, for example, more effective cooling methods, better soft magnetic materials, stronger permanent magnets, better electrical insulation, and alternative constructions. During the past few decades, investigating these aspects for wind turbine generators has led to significant progress (Fig. 5.1). Better arrangements of coils, magnets and soft magnetic materials enhance energy conversion. Direct-drive generators are built as ring generators with large diameters in different forms. When scaling them up, the support structure for the large diameter air gap becomes increasingly difficult.

These developments are expected to continue. However, after this long development time, the materials are used close to their physical limitations. For significantly shifting these physical limitations, better materials are necessary. Such materials that may lead to more disruptive technologies are being investigated.

Fig. 5.1 Stator of an electrical generator, designed specifically for wind turbines

Superconductors can carry current densities that are orders of magnitude higher than regular metal conductors. Therefore, they can break through the physical limitations of conventional machines. During the past decades, several prototypes of superconducting machines have been built. However, this has not yet led to the commercial application of superconducting machines. The integration of superconducting coils in a suitable cryostat and the simultaneous requirement for an inherently reliable cooling machine represents a huge challenge and necessitates an inter-disciplinary research effort. Evaluating the reliability of complete subsystems comprising superconducting coils, cryostat and cooling machine(s) requires extensive drivetrain testing facilities and a need for dedicated long-term experiments.

Basically, the same holds for carbon nanotube conductors, which are reported to be able to carry current densities that are much higher than regular metal conductors.

5.1.3 Electrical Conversion—Power Electronics

State-of-the-art power electronic converters for wind turbines are voltage source inverters using insulated gate bipolar transistors (IGBTs) or integrated gate-commutated thyristors (IGCTs) based on silicon (Si) technology as power semiconductors. The most important limiting components in these converters are the power semiconductors. They are basically characterised by the rated voltage levels, the rated current levels and the switching frequencies they can handle. Increasing the voltage level beyond rated leads to electrical breakthrough. Increasing the current level or the switching frequency beyond rated leads to overheating. If the required voltage or current levels are higher than rated, power semi-conductors can be connected in series or in parallel to obtain higher voltage or current levels and hence higher power levels.

Continuous research in power electronics has resulted in continuously increasing maximum voltage and current levels, maximum switching frequencies and the efficiencies of power electronics. Furthermore, there are alternatives for the state-of-the-art converters used today. These alternatives must be investigated because of their potential for offering superior efficiency or higher availability. The most promising solutions are as follows:

- switching to other power semiconductors based on silicon carbide (SiC) or gallium nitride (GaN);
- switching to alternative power electronics converter topologies, such as modular, multilevel, and/or fault tolerant technology;
- switching to other topologies at system level by better integrating power electronics and the wind farm collection system.

5.2 What Are the Failure Mechanisms and Is It Possible to Avoid or Mitigate Them?

Although combinations of gearboxes, electrical machines and power electronics have been used in wind turbines for decades and in other applications for over a century, their re-liability and availability are not at an acceptable level, especially for offshore wind. By looking at various recent failure analyses, it becomes clear that the gearbox, the generator and the electrical system are relevant drivers of the wind turbine downtime. Important questions are what the main causes for premature failures are, and whether and by what means these could be avoided. In order to address this matter, interdisciplinary research involving the manufacturers will be the key for improvement. As well as in other sectors, this will lead to increased reliability like in the automotive industry.

Furthermore, a better understanding of the physics of these failure mechanisms will probably also enable better condition monitoring, which consequently can be used for improving condition-based maintenance and condition-based control.

Scientific challenge 5-2: To develop a thorough understanding of the key failure mechanisms of electromechanical drivetrain components to design more reliable drivetrains and to develop better methods for condition monitoring.

5.2.1 Mechanical Conversion Systems—Gearboxes

The parts of the gearbox that fail most often are the roller bearings (two-thirds of the gearbox failures) and the teeth (about a quarter of the gearbox failures). Depending on the gear stage, these failures result in a non-negligible share of the wind turbine downtime. For example, a serious failure of the planetary stage can result in the need for a crane and disassembly of the gearbox on the ground.

The knowledge base on downtime causing failures is fortunately in an advanced state. However, the current understanding of some specific failure mechanisms and methods to avoid or mitigate the resulting failures is insufficient. For example, the failure mechanisms leading to, for example, WEC (white etching cracks) at the spur gear stages are not sufficiently understood. The same is true for failing bearings of the planetary stage, for example, because of moving bearing rings or unfavourable load distributions.

Modifying the bearing design (omitting bearing rings, electrically insulated bearings, black oxide-coated bearings, plain bearings) is a possible solution, yet system analyses of the failure mechanisms in real scale are still needed. Another example of a gearbox design change that is being investigated is the replacement of roller bearings with plain bearings.

Advanced simulation models and tools and extensive (accelerated lifetime) drivetrain testing are two common methodologies employed for investigating these

Fig. 5.2 Analysis of a gear box

complex issues. There is a need to validate the improved models of the effect or new designs and optimised control strategy on the local loads inside the gearbox (Fig. 5.2).

Some gearbox failures take time to develop, so it makes sense to monitor the condition and carry out condition-based maintenance if necessary. It is rather common to monitor the condition of gearboxes using vibration, oil property and particle contamination measurements.

5.2.2 Electromechanical Conversion—Generators

Although generators are a rather well-known technology, generators in wind turbines fail much more often than anticipated. What are the main contributing causes of premature wind turbine generator failures and could these be avoided by using superior designs? Although the number of failures in generators is low compared to power electronics, the generator availability must still be improved. Statistical data indicate that generators in wind turbines fail more often than comparable electrical machines in other applications, where a two-decade mean time between failures (MTBF) can be achieved. Generator failures are mainly due to bearing-related issues and winding short circuits. Possible causes for early failures include the impact of harsh environmental conditions of wind turbines, such as humidity, salt, and thermal cycling. Also, mechanical vibrations related to the complicated system dynamics may lead to premature failures.

It is rather common to monitor the generator temperature. Because most generator failures are bearing failures, vibration measurements also make sense. A better understanding of the failure mechanisms may disclose other effective methods of condition monitoring.

5.2.3 Electrical Conversion—Power Electronics

The most urgent issue with wind turbine power electronic converters is that they regularly fail. An essential research aim is therefore to gain insight into main causes leading to premature failures in these power electronic converters. An interlinked research question is what should be done to minimise and even avoid these premature failures by changing to better overall designs. Current knowledge about durable design of these systems is still insufficient. Contributing causes held to be responsible for early power-electronic failures include single and cumulative impacts linked to operating and environmental conditions like, in particular, humidity, salt settlement, vibrations, rapid temperature changes, and thermal cycling.

Failures in power electronics mostly develop very fast and come as unexpected. A better understanding of the failure mechanism physics may disclose early indicators for these failures. The voltages and the currents in power electronics are continuously measured by the power electronics control. These measurements contain an enormous amount of information about the condition of the wind turbine and the drivetrain components that is now seldom used.

5.3 How Can the Multi-scale Problem of the System Dynamics and Local Loads Be Evaluated?

How should the impact of wind turbine system influence and system dynamics be evaluated in order to identify design driving load cases or faults, especially when wind turbines are scaled up to sizes beyond what is currently available? This question has to be answered for all drivetrain components (gearbox, generator, power electronic converter) down to the individual machine element level to evaluate local loads.

This question is a multi-physics problem, because it involves different aspects, like aerodynamics, mechanics, electromagnetics, power semiconductor physics, and thermal dynamics, combined with the multi-scale problem of different timescales, which range from microseconds for the switching of the power electronic converter to seconds for the mechanical dynamics of the turbine and to minutes for the thermal effects. In some cases, the different physics can be studied separately. However, as an example, in direct drive generators, the coupling between dynamics of the mechanical construction and the dynamics of electromagnetics in the generator may be significant.

Finite-element models of a full energy conversion system are too "heavy" from a computational perspective to achieve a systematic, detailed investigation of the dynamics during different load conditions and under different fault conditions. Simplified representation, such as the combination of BEM codes with the structural parts of direct drive, becomes necessary. A method for transferring a finite-element representation or technical drawing into a BEM code has been proposed for development. Another solution would be to work further on local or sub-modelling with an increased attention to global specifications and boundary conditions.

An important aspect in specifying dynamic load conditions as well as fault conditions is to fully understand how specific faults could be mitigated from the electrical grid side to mechanical parts of the system and vice versa. Clear examples are low-voltage ride-through (LVRT) and failures in the power electronics causing short circuits in the generator. Hence, a great deal of effort is made to characterise such faults on system test benches; these make it possible to trace back the faults into a specific external fault condition for a given component such as the generator or gearbox. As an important first-step system, test benches will make it possible to learn about the relationship between the inner local (component) loads and the external loads and under what control conditions they occur, as these loads are not sufficiently understood. Elaborated measurement technology with the associated infrastructure is necessary to do that. System tests will have to be devised that provide new insight into the understanding of loads and wind turbine failures, as well as additional insight into future optimisation directions for all types of electromechanical conversion systems. In addition a holistic approach is essential in order to compare different conversion systems right from the beginning.

A follow-up question is whether the very fast control of the power electronics could be used for load reduction of the rest of the turbine during abnormal conditions such as grid voltage dips, loss of grid, and generator or power electronics failures. Because of the very fast control of the power electronics (of the order of ms), this could significantly influence these loads. However, the power electronics control can only influence these if such an abnormal condition can be detected sufficiently early and if proper measures can be taken.

Scientific challenge 5-3: To develop experimentally validated integrated design models and tools allowing the evaluation of the multi-scale problem of the system dynamics to improve understanding of local loads and to enable the design of more reliable electromechanical drivetrains.

5.4 What Are the Best System Choices?

The complexity of all the requirements, the different materials and their (mechanical, electrical, magnetic, thermal) limitations, and the different ways to build them into an integrated electromechanical conversion system make it difficult to decide what the best electromechanical conversion system is. It is a fact that there are a range of different commercially available generator systems, with international parties in addition working on alternative solutions. This may indicate a knowledge gap in determining the optimal conversion system.

Scientific challenge 5-4: Developing models that enable the design of the best electromechanical conversion system, given the complexity of all the requirements, the different materials and their limitations, and the different ways to build them into an integrated electromechanical conversion system.

5.4.1 Which Criteria Should Be Used to Evaluate Electromechanical Conversion Systems?

The state-of-the-art design optimisation criterion for wind turbines is the cost of energy (CoE). However, this is just an economic criterion. There may be other aspects (biological, juridical, aesthetical, political) that have a decisive influence on some of the choices that lead to a higher cost of energy, but possibly better solutions. As long as there is no better criterion, it makes sense to use CoE as a criterion.

5.4.2 Drivetrain Component Selection

CoE is the key criterion and key driver for comparing and optimising wind turbine drivetrains, including their power transmission systems, electrical power generation and conversion.

Evaluation of this criterion is a challenge. It includes (1) evaluation of the wind resource at a location using the Weibull distribution; (2) the cost of designing, manufacturing, installing, maintaining and dismantling wind farms and their civil and electrical infrastructure (cradle to cradle); (3) the efficiency and the availability of the wind farms; and (4) the cost of financing the wind farms. It should also include the cost of the consequences of integrating wind energy into the power system. It is quite a challenge to quantify all these in a proper way, especially because several of these factors vary at least with time, location, and political choices.

Although there are models for estimating the CoE, these models are generally not sufficiently detailed and are incomplete. They are not sufficiently detailed, for example, to calculate and compare the cost of maintenance for different electromechanical conversion systems. They are not complete because they do not include all the possible electromechanical conversion systems. A conversion system's total investment costs can be determined from individual materials'/components' cost by attaching a lifetime tag to each specific component. In a next step, maintenance and loss of availability costs associated with exchanging a component can be specified in order to estimate these cumulative costs. CoE is finally evaluated at system level by determining net energy yield, which is a function of gross energy yield minus production-related losses, set against the cumulative lifetime-based investment and operating costs.

For a reliable CoE estimate, specifying and validating component lifetime and establishing the main failure mechanisms are all crucial. The above methodology could be applied as a first step in an overall energy conversion system design phase followed by specifying drivetrain test conditions to be examined and verified experimentally at a drivetrain test bench. As the final process step a full-scale turbine test could be performed either in the field or on a system test bench.

5.4.3 Availability of Materials

The availability and cost development of materials to be used in future electromechanical conversion systems is a relevant issue. The price of commonly used materials may change significantly, as has happened for copper and, in a more extreme way, for rare earth element magnets. For new materials, like superconductors or wide-band gap power semiconductors, there may be a large potential for cost reduction.

Several manufacturers have introduced or are considering permanent magnet generators because of their improved efficiency especially during partial-load operation. The rare earth element material requirement and the cost of powerful (Nd, Pr, Dy) FeB magnets are highest in direct-drive generators. Because dysprosium (Dy) is the most problematic material, it makes sense to investigate the properties of rare earth magnets with a low amount of Dy. These magnets tend to have lower coercivity at higher temperatures, which makes them more sensitive to demagnetisation.

Chapter 6
Reliability and Uncertainty Modelling

G.J.W. van Bussel, J.D. Sørensen and P. Tavner

State of the art

The reliability of modern onshore wind turbines is at present at an acceptable level. This means that, assuming 100 % site accessibility and implementing a well-organised efficient maintenance strategy, most onshore wind turbines show availability levels of above 98 %. In other words the yearly downtime due to service and repair is less than 1 week.

Avoiding failures in wind energy systems is important because of the impact on downtime and thus on annual energy production (AEP), installation upkeep costs and hence CoE. This is especially valid for offshore wind power since the consequences of a failure offshore are more severe in terms of additional downtime due to limited accessibility, the complexity of offshore repairs and hence the overall production costs. Components of present wind turbines are as reliable as similar components in other applications and further increasing the reliability of components may very well lead to higher levelised cost of energy (LCoE). For designing an optimised system, the focus should therefore not be on significantly improving component reliability since that might imply a higher CAPEX (capital expenditure) and hence less favourable LCoE.

Instead, future R&D should focus on minimising and, if possible, even elimination of unexpected failures. In other words, the wind industry should move away from corrective maintenance (remedying of failures) and more towards preventive, predictive maintenance (repair just before failure) in order to achieve maximum availability.

G.J.W. van Bussel (✉)
Delft University of Technology, Delft, The Netherlands
e-mail: g.j.w.vanbussel@tudelft.nl

J.D. Sørensen
Aalborg University, Aalborg, Denmark

P. Tavner
Durham University, Durham, UK

G. van Kuik and J. Peinke (eds.), *Long-term Research Challenges in Wind Energy - A Research Agenda by the European Academy of Wind Energy*, Research Topics in Wind Energy 6,
DOI 10.1007/978-3-319-46919-5_6

An additional approach is to incorporate/activate system redundancies, whenever possible and technically and economically feasible, in order to keep the wind turbine and/or wind farm in operation after a failure. This should only be implemented when a reduction in LCoE can be achieved, and would allow fault-tolerant operation to minimise downtime. A further additional approach is wind turbine and wind farm uncertainty modelling, which is still in its infancy to date but forms an intrinsic aspect for minimising unexpected failures and downtime. Finally, research methods and tools should aim at obtaining a more uniform reliability level for components and subsystems, where uniform is to be understood as uniform from a risk point of view, i.e. also accounting for the consequences of failure.

6.1 How Can Operation and Maintenance Data Be Made Available to Increase O&M Knowledge?

Data collection on wind turbine failures and O&M activities has become standard practice, but it is still not standardised, often performed by wind turbine manufacturers and typically not publically available.

Maintenance-related knowledge together with sensor signal information[1] and feedback of this knowledge into the loop for turbine optimisation could lead to improved performance and reduced costs. However, both an overall methodology for documentation and a suitable model for analysis are lacking, thereby prohibiting the development of a sound scientific basis.

Scientific challenge 6-1: To set up a framework in which confidential data and documentation can be anonymised and processed for serving the need of defining historical and the state of the art in wind turbine and wind farm reliability data, and O&M demand in a well-structured fashion.

6.1.1 The Initiation and Maintenance of a Comprehensive, Representative and Publicly Accessible Turbine Upkeep and Failure Database

Today's most cited database is an outcome of a public German project. The bulk of the data was already collected in the 1990s and thus refers to wind turbine technologies very different from the current state of the art.

One main achievement of a European project concluded in 2011 was the determination of a taxonomy terminology and data collection standard, but this

[1]'Sensor signal information' is the indication for the data collected by specific sensors, other than the usual SCADA data collection.

taxonomy did not comply with the current international designation system for power plants. New activities on the initiation of a public database and on taxonomy unification are being developed by the International Energy Agency.

With the key elements for initiating a public database available, a next pending and urgent challenge will be to initiate and maintain such an anonymised database. Despite the fact that this is not a purely scientific issue, the topic remains of key importance for academic, engineering, manufacturing and finance sectors of the wind industry. Achieving this requires support from all these sectors and serves a common goal towards more reliable and more economical wind power generation.

6.1.2 The Documentation of Service Visits and Maintenance Operations

Substantial added value can be generated on maintenance-related knowledge through planning sound and systematic inspection and reporting. Together with sensor signal information and development of a physical systems model such fresh knowledge could be fed back into the model loop for turbine upkeep optimisation. However, both a suitable model and an overall methodology are lacking, as well as models and tools for supporting a decision-making process aimed at optimised upkeep planning (Fig. 6.1).

6.2 How Are Data Selected, Collected and Analysed, and How Can They Be Used for Improved Performance?

In order to answer these questions it will be necessary to know in sufficient detail about the external conditions today's wind turbines currently experience and how this could change in future. It is also necessary to exploit current monitoring systems to the highest extent and introduce new hardware and software tools to optimise operational performance.

Scientific challenge 6-2: To obtain sufficient knowledge and information to optimise the operational performances of present and future wind turbines and wind farms. Extensive knowledge of external conditions, sensor signal information, structural degradation models of systems and components, and probabilistic models for deterioration assessment is needed.

Fig. 6.1 Implementation of maintenance strategies requires courage, in particular at offshore sites

6.2.1 Knowledge About External Conditions

With respect to external conditions—i.e. wind speed and direction, wind shear, turbulence, wind veer, waves, currents, humidity, temperature, ice accretion and wake operation characterisation—knowledge of such conditions, often beyond current IEC class standards, is essential. Reliability degradation is often related to complicated temporal combinations and sequences of external conditions, control strategy and internal loads. But what are then the key features required of these external conditions for enabling more advanced forecasting with regard to degradation and faults, and how could these external conditions be determined?

6.2.2 SCADA and Condition Monitoring Signal Retrieval and Analysis

SCADA (supervisory control and data acquisition) signals, available in all modern wind turbines, are today seldom applied for turbine health monitoring. A lot more useful information can be potentially extracted from SCADA together with increasingly common condition monitoring system (CMS) signals. Properly integrated analysis of combined SCADA and CMS signals might result in superior assessment of an asset's actual condition state in terms of operational health, degradation damage and remaining lifetime. Unfortunately, access to SCADA data sets is often proprietary. When such access is granted, is the scientific community then sufficiently equipped with the right tools for conducting such analysis? Finally, should model-based reasoning techniques and tools be developed to facilitate this process?

6.2.3 Improved Maintenance Methods

Service life design assessment aims at accurately determining the (remaining) life of a component. Initial lifetime assessment can be performed on the basis of physical design and design specifications, plus structural (physical) modelling. Remaining component or system service life calculation should be performed through sensor signals analysis and systematically guided inspection following a prescribed methodology using a standardised reporting structure with automated reporting tools.

Condition-based operation is sometimes proposed as a smart maintenance tool, aimed at intelligent upkeep and/or fault-tolerant control, but is rarely used. The key issue is to maintain operation and hence power production even if certain components are close to failure or have failed already. In the first case, operational modes are modified to extend operating time until a suitable remedy is found. In the second case the system is reconfigured to a basic function whereby power generation can still be maintained.

The targeted goal is to optimise upkeep to maximise annual energy production and minimise operating expenditures (OPEX). This could be achieved by postponing major component exchange up to the moment before actual failure, and by assigning the preventive component exchange to a suitable time slot.

6.2.4 Implementing Mathematical Probabilistic Approaches to Improve Operational Performance

Bayesian updating, partially observable Markov decision processes and other mathematical maintenance optimisation and decision-making tools have been developed in probabilistic design and risk-based decision making. Implementing such tools in the wind energy field is currently a new, actively pursued research area for which initial efforts have already shown promising results. Such approaches are only possible if physical inspection of components and/or physical assessment of the state of deterioration is possible and can become important when performed as part of the operational monitoring process. The physical information can be either direct or indirect though indicators.

6.3 What Are the Best-Suited Stochastic Models and Parameters for Performing Adequate and Accurate Systems Reliability Assessments?

Different mathematical models for systems reliability modelling are currently used to describe the reliability of wind turbine components and systems. The presently implemented systems reliability approaches can be adopted for O&M optimisation. This would support the identification of main areas important for tools and code development. This also allows for definition of the most important areas for improved design and extended operational time, and assessment of the impact on LCoE and AEP.

Scientific challenge 6-3: To determine adequate stochastic models and parameters suitable for wind turbine and wind farm operational modelling. Risk levels need to be determined in order to facilitate probabilistic decision-making processes in the operational phase of wind turbines and wind farms.

6.3.1 Stochastic Models for Expressing Reliability and Failure Rate Characteristics

At present both exponential, lognormal, normal, and Weibull are all being used for describing failure probabilities. But which of these distributions is most appropriate for given specific failure types is not clear. A well-known methodology for describing failure patterns and rates is the bathtub curve, which describes changes in component or system failure rate on a timescale. But what are valid β values to describe the three operating phases, split into early, regular and wear-out, for different components or systems?

A complicating factor is that external conditions play a much larger role in wind turbine failure accumulation and failure degradation compared to other generation systems, and these external conditions have their own stochastic conditions to be considered as well.

6.3.2 Simulation Models for Downtime, Availability and Annual Energy Production Calculation

An unanswered question is, what are the best simulation methods for determining downtime, availability and annual energy production? Deterministic approaches, Markov chain modelling and Monte Carlo simulations are all in active use, but are these sufficiently adequate for providing accurate forecasts?

Availability is one of the key performance indicators for optimal O&M/upkeep approaches. Implementing smart maintenance and fault-tolerant control options might enable such an optimised strategy since it would allow continued operation even when a main component has failed. However, no methodology exists as yet to implement a design-for-availability strategy. On the other hand, risk-based maintenance is already applied in many high-tech areas and could prove a starting point for systematic design-related availability assessments in wind energy technology. A key aspect for such a design-for-availability approach would be the implementation of decision-making processes based upon probabilistic modelling.

6.3.3 "External" Information About Anticipated Risk Levels and Quality Assurance and Control

A required lead for an availability strategy for design would be the use of "external" information/knowledge as building blocks for the implementation phase. These building blocks would include a determination of anticipated risk levels, onshore and offshore wind systems requiring different approaches, and higher quality assurance and control (QA&C) standards in both design and manufacture.

Fig. 6.2 Training of personnel on how to act under emergency situations

Standards for risk are not yet generally set and will differ from operator to operator. QA&C standards are in place and implemented today in wind turbine design and manufacture, but they are not yet fully developed or deployed for maintenance and repair (Fig. 6.2).

6.4 When and Where Are Detailed Physical Models Needed to Predict Degradation and Failure?

Knowledge about the modelling physics of degradation and failure is now available for a wide range of technologies. However, the required level of accuracy strongly depends upon specific usage. A first topic is to identify for which turbine components and (sub)assemblies detailed physical models are needed. Once identified the question is whether it will be possible to determine the accuracy levels required in structural reliability engineering. Evidently, any level of accuracy for wind energy use depends upon the levels achieved in other application areas and is interlinked to structural reliability engineering. This includes for instance specific knowledge on accuracies required in material properties and manufacturing tolerances. This research question is hence also highly relevant for the materials and structures area of R&D (see Chap. 1).

Scientific challenge 6-4: To strongly improve the structural failure modelling and to take initiatives to adapt structural failure models used in other areas of technology. To acquire detailed information about the design and the operational conditions of wind turbines. To

define future levels of design, manufacturing, material and maintenance tolerances. To acquire more extensive information about the actual external conditions in which wind turbines and wind systems are operating.

6.4.1 Degradation and Failure Modelling

A key aspect of fault forecasting is the full understanding of the physics behind degradation, the process of deterioration and the implementation of that understanding in the design stage, known as damage/deterioration modelling. An important aspect here is the influence of external conditions on degradation of components and (sub)assemblies. Not only do these need to be determined upfront, they also need to be monitored during operation.

In the operational phase, well-structured and intensified monitoring and inspection processes might contribute to updating expertise and insight into the remaining lifetime of crucial components. This, in turn, is essential for further optimising O&M planning and execution focused on reducing downtime as well as lowering operating costs.

These are all considered key elements for a (future) research programme and should be carefully matched and fine- tuned with the materials and structures area of R&D.

6.4.2 Tolerances in Material Properties, Manufacturing, Assembly, Components, and/or Full Installation Exchange Inaccuracies

Physics modelling is not the only important factor in fault forecasting; in addition, knowledge and insight about deviations in material properties, manufacturing tolerances and inaccuracies that inevitably occur during assembly and installation are also important.

The key question here is whether these tolerances in material properties, manufacturing, assembly, components, and/or full installation exchange inaccuracies are sufficiently known and documented. Addressing these issues is probably more an area specifically suited for certification bodies instead of the research community. Despite this observation, the specific knowledge field itself remains of key importance for pursuing further research focused at optimising systems structural reliability. One of the main future research goals is achieving a superior capability to determine and analyse faults and fault frequencies.

Chapter 7
Design Methods

C.L. Bottasso and M. Muskulus

State of the art

In this document, "design" is considered to include the choice of a wind turbine configuration and the sizing of the machine and all of its principal components, including rotor, drivetrain, nacelle, tower and support structures, as well as the choice of the generator, pitch and yaw systems. The design of a wind turbine implies the evaluation of the aerodynamic performance, loads, dynamic response, and stability in normal and extreme operating conditions, including the occurrence of faults. The design problem does not end at the level of the single wind turbine but expands to also include the design of the complete wind power plant, where the machines operate together and interact among themselves and with the environment. The disciplines involved in these complex processes include aerodynamics and flow physics, structures and materials, dynamics, and controls, as well as hydrodynamics in the case of offshore applications. For each of these disciplines, knowledge and design models have been and still are being developed as discussed in this document. However, these models have reached different levels of fidelity, applicability and computational costs, and cannot always be combined. The design of a wind turbine and of a wind power plant is therefore characterised by a marked multidisciplinarity with strong couplings among the various disciplines covered by models which may be different in complexity and proven validity.

C.L. Bottasso (✉)
Technische Universität München, Munich, Germany
e-mail: carlo.bottasso@tum.de

C.L. Bottasso
Politecnico di Milano, Milan, Italy

M. Muskulus
Norwegian University of Science and Technology, Trondheim, Norway

G. van Kuik and J. Peinke (eds.), *Long-term Research Challenges in Wind Energy - A Research Agenda by the European Academy of Wind Energy*, Research Topics in Wind Energy 6,
DOI 10.1007/978-3-319-46919-5_7

Although the industry has been developing wind power technology for decades, there are specific scientific barriers that should be removed and many areas where improvements could be made, leading to significant effects on the cost of energy, on the social acceptability, and on the environmental and human impacts of wind power.

7.1 How Can We Develop Truly Holistic Design Tools?

At the moment, a truly integrated approach to wind turbine design is still lacking. The design process is often based on iterations among discipline-oriented groups of specialists: each group addresses a sub-discipline (for example, aerodynamics, structures, controls, systems) and proposes solutions within that sub-discipline, which are then passed on to the next group. Iterations are conducted within and among groups until a best compromise solution is identified that satisfies all requirements and desired constraints. Automated algorithms often mimic such a manual procedure. Limits of such an approach include a possible lengthy and slow convergence towards satisfactory solutions, a possible incomplete exploration of the design space and lack of understanding of its characteristics, with possible trapping in sub-optimal choices. Similarly, tools for the optimisation of wind power plants are still in their infancy, and more advanced approaches are needed to deal with complex issues such as wake interactions and orographic effects, etc.

Scientific challenge 7-1: To develop holistic automated and comprehensive design methods for wind turbines and wind power plants for exploring the available design space and identify optimal trade-offs.

7.1.1 Holistic Multidisciplinary Optimisation by Physics-Based Modelling

One key question is how to formulate an integrated preliminary sizing of all major components in a wind turbine. In fact, the formulation of an integrated optimisation problem is a challenge, and multiple issues need to be addressed. First, the machine needs to be simulated in a large number of dynamic operating conditions. The stochastic effects of turbulence imply the need to collect sufficient samples in order to determine statistically relevant performance indicators, which often require significant computational efforts. Second, strong couplings among disciplines need to be considered. For example, choices in the aerodynamic shape of the blade affect not only performance and hence power production but also loads and the sizing of structural components, choices in the control laws affect performance, loads and duty cycles of the actuators, etc. Iterations among disciplines or approximations in the formulation and solution of the design problem may affect the way such couplings are considered, leading to sub-optimal choices. Third, the solution space may be rough, which significantly complicates the numerical solution of the optimisation problem. For example, changes in the control design may suddenly switch the dominating load of a given component from one design load case to another. When this happens, the dominating load may suddenly jump from one value to another, which indicates a lack of smoothness in the solution space. Such roughness of the problem will typically not be manageable with gradient-based solution techniques. Fourth, there might be multiple local solutions. Incomplete searches of the solution space might fail to identify global optima, remaining trapped in local sub-optimal choices.

Another important question regards the operational scenarios that should be considered within the design process. The choice and definition of the operating scenarios used during design need to be capable of faithfully representing the conditions that will be encountered by the machine through-out its lifetime. Although design and certification guidelines are available and continuously updated, progress remains to be made on multiple fronts, including the determination of extreme events, the effects of operation within wind farms or in complex terrains, the operation in offshore deep-water environments, the possible occurrence of faults at the electrical and/or mechanical levels, and the effects of software errors. All such operational conditions can have very significant effects on the design, and hence they play a crucial role (Fig. 7.1).

7.1.2 Uncertainties, and Probability of Failures

Yet another crucial question is how to include uncertainties in the design process. As for most engineering design problems, robust solutions are often more valuable than very sharp optima, where even minor changes in the conditions or parameters may lead to significant drops in the cost or violation of the constraints. Therefore,

Fig. 7.1 Overview of aspects to be taken into account when designing an offshore wind farm. Cover of the ECN Wind Energy R&D programme 2015

there is a need to include robustness and reliability aspects in the design process, so as to hedge against unavoidable uncertainties. Robust optimisation methods have been studied in neighbouring technical domains, and approaches are being developed to try to quantify and use information about uncertainty in models, loads, materials, etc. Reliability-based design optimisation methods have also been developed that use the probability of failure as a constraint or cost term. Such methods could be profitably applied and tailored to the specificities of the wind energy design problems, with possible improvements with respect to the currently adopted deterministic approaches.

7.1.3 Beyond the Single Wind Turbine

Wind farms are complex systems, in turn made of complex subsystems (the individual wind turbines), which interact among themselves and with the environment in very complex and yet not fully understood ways. Better knowledge is necessary in multiple areas, including wakes, interactions of the wind farm with the atmospheric boundary layer, wind farm internal dynamics, turbulence, propagation of disturbances, and the effects of controls on wake shape and characteristics. These topics are intimately linked with other R&D areas, such as aerodynamics and controls, but also have possible important implications in design, both of wind farms and the individual machines. For these reasons, a strong connection and coordination among these neighbouring fields should be among the priorities of future research activities.

What are the effects of wake interactions and internal wind farm turbulence and shear on the loading conditions of wind turbines? Are these conditions correctly captured by currently adopted design and certification guidelines? Is there any possible gain to be obtained by adopting a more wind farm-centred approach to design, rather than the classical approach where each machine is designed as if it were operating in isolation? Would machines designed from the very beginning to be part of wind farms look different from current wind turbines, and if so, what would be the main effects of such a change of design philosophy? Similar questions apply to the problem of operation in complex terrains. Here again, it is interesting to explore possible specific designs that might be better tailored and adapted to the peculiar characteristics of the flow conditions encountered in complex terrain environments.

The design of wind farms is often seen as a separate and distinct problem from the design of wind turbines. In fact, one first typically designs a wind turbine by treating it as a single operating unit and then designs a wind farm by selecting the most appropriate wind turbine model among those that are available on the market. Is this the best possible approach? This is certainly questionable, and it might be argued that different wind turbine and wind farm designs could be achieved by adopting a more comprehensive approach. For example, would it be possible to more densely pack wind turbines within a wind farm, to reduce land use and cost of connections, for example by developing specially adapted rotors and suitable control strategies that minimise interactions and promote faster wake recovery? To answer these questions, a greater integration among wind farm and wind turbine design tools should be pursued. This might possibly require the development of modifications to the specific modules and algorithms of the individual design tools.

7.2 What Are the Validated Simulation Models that Can Support a Holistic Multidisciplinary Design Activity?

Design is enabled by mathematical models and numerical methods, which provide the simulation tools used for assessing all design relevant quantities and effects. Clearly, the fidelity of such tools to reality is of crucial importance, as a design can only be as good as the simulation models that have been used to produce it. The fidelity of simulation models to reality depends on two aspects: their ability to capture the relevant physics and the correct tuning of their parameters. The former problem can be addressed by selecting or developing appropriate models; in this process, a balance is often needed between model complexity and computational cost, as excessively expensive models might not be usable within a design loop. The latter problem can be addressed by using calibration techniques that tune the model parameters to deliver the best possible matching with available experimental data.

Scientific challenge 7-2: To develop a full range of validated models of varying fidelity and complexity that can support all phases of the design of wind turbine and wind power plants, with a sufficient level of fidelity of all relevant physical processes, at reasonable computational costs.

7.2.1 Validated Multiple-Fidelity Simulation Models

For each discipline, suitable mathematical models and associated numerical algorithms are needed. Each needs to deliver a sufficient level of fidelity and be capable of achieving a sufficient accuracy for that specific phase of the design.

One important research objective would be the availability of a complete hierarchy of models, with different levels of complexity and hence cost, to support all phases of the design process. In fact, during design one typically needs to simulate physical processes at different levels of fidelity: lower-fidelity tools can be used for fast extensive explorations of the design space, while higher-fidelity ones are needed to refine the design in promising local solution pockets. Therefore, for each discipline a full hierarchy of models is needed, rather than a single one. Consistent connections among models of different fidelity are needed. These may be in the form of model compression techniques, which may extract reduced-order models from higher-fidelity ones (e.g. from 3-D FEM to beam models using cross-sectional analysis techniques). When direct reduction is not possible, connections and transitions among existing models should still be improved, which may lead to a better understanding of the limits and use of the various models. These topics are also relevant for control applications, as many modern control approaches are model-based. Therefore, the derivation of good-quality reduced-order models is also particularly important for the synthesis of advanced control laws, which, in turn, play a central role in design.

Validation of all such models is a crucial step, as validation and verification demonstrates whether or not a given simulation tool is capable of representing some physical processes to a certain level of fidelity. Validation relies on the availability of experimental data, which in some cases is particularly difficult to obtain. A mix of laboratory and field testing activities are needed to cover all needs of this very vast and challenging task.

System identification and parameter estimation theories have been developed for the purpose of tuning models to experimental measurements. These techniques have so far received relatively modest attention in the wind energy sector. However, as models of increasing levels of sophistication and complexity are being developed at a fast pace, a parallel effort in model calibration should be pursued. This is all the more relevant for design, as the quality of design is intimately connected to the quality of simulation models.

In support and integration of the more classical approaches described above, there are possible significant developments in the frequency domain analysis and other approaches, such as probability-evolution methods, statistical meta-modelling, or Bayesian load analysis. These models are still in their infancy, and much remains to be done. Of particular interest is the problem of efficiently and accurately assessing fatigue damage, where probabilistic approaches could bring substantial benefit.

Yet another crucial issue with modelling is related to nonlinearities. As an example, the determination of aeroelastic stability margins depends on the actual coupled (often non-linear) interplay among the structural, aerodynamic and control fields. The current models used for design purposes have, in general, a limited ability to capture nonlinear phenomena such as flutter, limit cycle oscillations, saturation effects, dynamical buckling, and many others. With the increasing size of wind turbines, these phenomena become more relevant, a problem which clearly calls for increasing attention to fully coupled models capable of capturing all nonlinearities of interest to a sufficient degree of fidelity to reality.

Fig. 7.2 Transport and installation loads are to be included in the design

7.3 How Are Cost Models Formulated so as to Capture the Effects of Each Technology and Design Choice?

Design can play a leading role in the development of wind energy. In fact, new technologies and ideas are being proposed at an ever-increasing pace. Through design, one can assess whether the benefits brought by a new technology (for example, load mitigation, improved power capture) are offset or not by its unavoidable cons (for example, increased manufacturing or operation and maintenance costs, decreased availability). Therefore, it is only when a technology is assessed at the design level that one can say whether it may be actually relevant or not. However, implementing such a design-centred point of view requires reliable cost models which can capture all relevant effects of a technology (Fig. 7.2).

Scientific challenge 7-3: To develop methods to assess technological innovations with respect to their design implications, enabling and accelerating future wind power technology development.

7.3.1 How Can Reliable Cost Models Be Established?

An important question is whether physics-based cost models could be formulated to a level of sufficient fidelity, such that they are usable as figures of merit for design.

The leading criterion for comparing design solutions is cost of energy (CoE) performance. However, complete and reliable cost models are currently not available, especially outside of the industry. Furthermore, several of the few cost models available in the public literature are based on historical data regressions. These have proven unreliable when new technologies not contemplated in these regression data points are being explored. Therefore, physics-based models based on first principles are needed. In addition, very often one is not only interested in one single solution that minimises a given cost but also in understating the trade-offs among different cost metrics. In all those cases, it important to be able to compute efficiently Pareto solutions that can help the analyst make the correct choices.

7.3.2 Design-Driven Technological Assessment

Any new technology will come with its pros and cons, as it will give benefits on some performance indicator but will invariably induce some cost to be paid on other figures of merit. By adopting a design-centred point of view, the trade-offs implied by any new technology can be assessed fully, accounting for its positive and negative impacts on all items of interest, including the CoE and effects on the population and the environment. Although it may be sometimes difficult to draw conclusions of general validity, as the result of the assessment will often be specific, the use of a design-centred approach further stresses the need for effective design tools and reliable cost models.

Control is a good example of a technology that may benefit from a design-centred paradigm. In fact, control is often seen as a separate discipline: new methods and algorithms are proposed and then applied to existing machines to evaluate the effects on performance, fatigue, extreme loads, vibrations, etc. A design-centred approach to control may be more effective. In fact, by adopting a design view, one may realise what conditions drive the sizing of each wind turbine component. For example, some components might be driven by ultimate loads encountered during shutdowns in extreme gusts, so that their reduction could be performed by optimising the way this manoeuvre is conducted. Other components might be driven by storm conditions when the machine is idling; in such a case, load mitigations can only be achieved by changes in the rotor design and certainly not by improvements in the controls. Without taking this view, one might work on control solutions that reduce loads that, not being design drivers, have in the end no or a very limited impact on the optimisation of the machine. Furthermore, the analysis of load rankings gives useful indications on how much gain one should seek from a new advanced controller. For example, assume that a dominating load can be reduced by improving the control logic, and that this load is 10 % higher than the next one, which happens to be due to—for the sake of an argument—storm conditions. Then there is no need to reduce this load more than 10 %, as further gains would not give any benefit from the point of view of design. Such a design-centred view, routinely adopted by industry, could be more often adopted in the scientific assessment of new methods in order to better understand the true impact that advanced controls may or may not bring to wind energy technology.

Chapter 8
Hydrodynamics, Soil Characteristics and Floating Wind Turbines

M. Muskulus and D. Matha

State of the art

Support structures for offshore wind turbines fall into two distinct types. Bottom-fixed turbines are connected to the seabed in a manner that restricts vertical motion; key examples of this class of concepts include monopiles, gravity-based structures, or jackets. Commonly, a transition piece connects the turbine tower and support structure. For monopiles this traditionally involves a grouted connection, but grout failure issues have recently led to a revision of the design practice and standards. A coning shape of the matching surfaces with or without shear keys is now increasingly employed for securing the transition piece and an overall higher quality of the grout connection. The seabed connection can be in the form of hammered or drilled piles, or suction buckets. Alternatively and with gravity-based solutions the mass of the support structure can be used for a stable and secure structure-seabed interface. Offshore bottom-fixed turbines have been realised routinely in commercial wind projects since the late 1990s, but the current market situation is characterised by continuously increasing water depths and a trend towards bigger turbines. This in turn leads to a constant shift away from the current state-of-the-art and to increasingly larger support structures.

Limiting factors are manufacturing constraints as well as an increase in hydrodynamic loads and the resulting dynamic response experienced by the structural members of larger support structures with increasing diameters. Uncertainties with respect to soil properties and time-consuming installation processes are also a defining factor.

With monopiles, the current state of the art lies at about 30–35 m water depth requiring structures of 8–9 m outer diameter. As an alternative, jacket-type

M. Muskulus (✉)
Norwegian University of Science and Technology, Trondheim, Norway
e-mail: michael.muskulus@ntnu.no

D. Matha
University of Stuttgart, Stuttgart, Germany

G. van Kuik and J. Peinke (eds.), *Long-term Research Challenges in Wind Energy - A Research Agenda by the European Academy of Wind Energy*, Research Topics in Wind Energy 6,
DOI 10.1007/978-3-319-46919-5_8

foundations that are built like an open space frame structure and therefore more transparent to waves have been explored since 2006. These structures can be favourably employed up until at least 50–60 m water depths.

The main limitation with support structures is often not technological but first and foremost economic in nature. For instance, even though it might be technically feasible to design and install larger-diameter monopiles or deep-water jackets, this might prove too expensive compared to other support structure designs.

For deep water in excess of 50–60 m, floating turbines or "floaters" become increasingly attractive alternatives to bottom-fixed turbines. Full-scale floaters have only been realised in the past few years and so far only as demonstrator projects. Especially with floating spar-type turbines (i.e. Hywind) and semi-submersibles (i.e. WindFloat) some years of experience have been gained.

Whereas a significant amount of valuable experience has been transferred from the oil and gas industry, it is being realised that conditions for offshore wind turbines differ significantly. Support structures in the wind industry are smaller and the overall conditions more dynamic, with considerable consequences for the design process. For instance, whereas frequency domain analysis is in many cases useful and sufficiently accurate for the design of offshore platforms, the same method cannot be employed one to one for offshore turbines. This is due to non-linearities both in the rotor system and the bottom-fixed soil model or floating turbine mooring system. Regarding time-domain simulations, the state of the art is defined by current tools, which are still limited in terms of features.

8.1 How Can We Predict Hydrodynamic Effects Relevant for Offshore Wind Turbines More Efficiently and Reliably?

Hydrodynamic phenomena are complex and manifold, and include the known difficulties of both free-surface flow and turbulence. A first consequence is that both Reynolds number and Froude number are important for structures subject to hydrodynamic loads. Simultaneous scaling of physical models therefore becomes challenging, if not impossible, especially if a wind turbine is also to be included. Regarding numerical models, the high Reynolds number in marine conditions (water) means that the fluid dynamics can often be approximated by potential flow. Normally it is necessary to supplement these models with information about the behaviour of a structure in real waves, for instance by measuring drag and damping coefficients through physical wave basin model tests (Fig. 8.1). However, this is both time-consuming and expensive.

Scientific challenge 8-1: To be able to predict the complex fluid phenomena of wave structure interaction for an offshore wind turbine on first principles.

Fig. 8.1 Breaking waves contribute significantly to the mechanical loading of an offshore wind turbine structure

8.1.1 Life Beyond Potential Flow and Linear Wave Theory

Potential flow calculations currently lie at the heart of motion response analysis for floating structures. Potential flow is based on the assumption of inviscid, incompressible and irrotational flow, assumptions which are not sufficient to capture all relevant load effects acting on the often complex floater geometries. This implies the question of how additional viscous effects can be accurately calculated and predicted. An important case in point is the oscillation of a floater with heave plates: can the viscous damping in such a system, primarily caused by vortex shedding, be predicted reliably from first principles?

Waves in general are nonlinear phenomena, and even the simplest wave load models incorporate a quadratic nonlinearity in water particle velocities. More accurate wave theories and load models result in higher-order nonlinear terms. To what extent are these nonlinearities relevant for the dynamics and the design of offshore wind turbines? And, more particularly, under what conditions do second- and higher-order hydrodynamic effects have to be considered, and when will it be necessary to apply the relative formulation of the Morison formula? How does the existence of currents modify these results, and how should corrections such as stretching to the instantaneous free surface be implemented?

Breaking waves are considered a particular case of a non-linear wave phenomenon. What are the main impact characteristics of breaking waves on multi-member structures like offshore wind turbine jackets, both in terms of the wave loads' time course and their statistical description? And how do breaking wave loads depend on seabed slope, wave climate and other defining factors? With what likelihood do breaking wave impacts and other extreme wave events occur for a given location offshore?

Connected to breaking and steep nonlinear waves is the phenomenon of wave run-up, wave slapping and slamming loads and green water loading. These effects cannot be considered by potential-flow-based methods, and currently only CFD computations are able to model such effects. How can the effects of wave run-up, wave slapping and slamming loads and green water loading be considered in efficient simplified methods or estimated from first principles?

A mooring system provides station keeping for floating structures and restricts horizontal motions. In the case of a catenary mooring, the mooring line dynamics have to be taken into account from a certain cable length onwards. Will it be possible to calculate or estimate the amount of mooring line damping from first principles? What are reliable and efficient prediction methods for flow-induced vibrations of mooring lines and power cables?

A number of regular nonlinear wave theories do exist, but irregular waves are based on spectral synthesis and the superposition principle of linear wave theory. A question is whether there could be a useful concept for random nonlinear water waves.

Is a numerical wave tank, such as a CFD simulation of water waves, useful or even necessary to capture some hydrodynamic effects that cannot be described otherwise or predicted efficiently? Can it ultimately replace a model test, as is the case in aeronautics? Are vortex methods or smoothed particle hydrodynamics viable approaches?

8.1.2 Numerical Tools and Validation

There is also a general need for more realistic combined wave condition and wind pattern testing conditions in order to increase the information gained from these load cases while not significantly increasing the number of simulations needed. Currently only a limited number of sufficiently general simulation tools for a bottom-fixed support structure, such as a detailed offshore turbine jacket model, are available, while existing tools are too slow. With most tools, key built-in capabilities such as code checks and an efficient workflow are also missing.

Available tools have only been validated with limited real-world measurements on a statistical basis. No full-scale validation was performed for the detailed time-domain dynamic response of the structure subject to combined wind, wave and current loading. There is a further need for validation protocols and novel strategies, and the issue of updating simulation models when the dynamics do not match measurements is a largely unexplored research field. Validation with physical models needs special attention, since, for instance, wind conditions must be more realistic than currently possible, and scaling issues arise. An interesting possibility is offered by in-the-loop approaches whereby part of a model is simulated. Current simulations aim at being as realistic as possible by including all physical effects that might be relevant. However, there is also a need for developing robust and efficient simplified models that can be used within optimisation algorithms and model-predictive control strategies and which can be linearised for control design purposes. This need is additionally warranted from the perspective of hiding design data in collaborative industry projects. Blackbox approaches to support structure dynamics based on impulse-response functions or statistical meta-models incorporate this built-in capability.

A defining marine environment main characteristic is that structures are always subject to excitations. Techniques for structural health monitoring, vibration and data analysis must be capable of coping with such ambient excitations. As the input is typically not known, a normal input–output formalism cannot be used and special statistical approaches must be employed. In the few cases that high-fidelity time-resolved data on environmental conditions are available, for example using wave radar, it has proved immensely difficult to match observed and simulated structural behaviour. There are also very few observations of extreme events, resulting in large uncertainties. Another open challenge is how to deal with fatigue for concrete structures in seawater. In general, all big challenges that have not been solved in the past 30 years represent a big risk as well as offering huge potential gains.

8.2 How Can We Efficiently Characterise Complex Properties and Highly Nontrivial Response of Soil Characteristics?

Soil–structure interaction is often perceived as rather complex and is therefore neglected or only considered in a highly simplified form. However, the topic has in the past years received a lot of attention after failures in grouted monopile connections occurred. It is further hypothesised that soil dynamics could contribute to a substantial part of total damping in wind turbines. However, a limiting factor is that current dimensioning methods and tools are based on data obtained for smaller piles than used today in offshore wind. Numerical simulation with solid soil elements, on the other hand, is highly complex and unfeasible on a large scale.

Scientific challenge 8-2: To understand the implications for wind turbine response and design loads that are based on the complex soil structure interaction.

8.2.1 Models for Soil-Structure-Interaction

Soil is a highly complex material. Current state-of-the-art pile sizing and simulation models for soil–structure interaction are based on the so-called *p–y* method. This model has been developed from data obtained for different pile sizes and a re-evaluation is due. A main question is whether there is an efficient alternative to *p–y* curves. Another regards generalisation of *p–y* curves, like dynamic *p–y* models that utilise different near- and far-field properties, and whether this would allow for more accurate load simulations, and, furthermore, in what way parameters should be determined. Novel foundation concepts such as suction buckets pose additional challenges and it is unclear how to simplify the soil–structure interaction such that it can be efficiently implemented in numerical wind turbine simulation models.

A complication is that soil models for piles are typically 30–50 years old and very conservative, and soil properties are in general also history-dependent. Cyclic stresses are known to result in systematic soil stiffness degradation. A main question is how such cyclic soil degradation effects can be effectively described and incorporated into existing simulation models. A second question is how these effects accumulate over a 20-year design life.

Furthermore, what is the amount of soil damping for an offshore turbine? Is it possible to estimate soil damping from first principles, like from numerical simulation with solid elements? Improved insight could also lead to major breakthroughs like a possible pile eigenfrequency fine tuning through varying ramming depth as a function of soil characteristics and other key variables.

8.2.2 Uncertainties in Geotechnical Design

As it is currently not possible to reliably predict soil conditions for each potential wind turbine position in a planned wind farm, it is essential to know the degree of uncertainty in field measurements and soil studies and how to best deal with soil-related uncertainty in load simulations during an offshore turbine design phase. With what kind of uncertainty does one have to design the turbines, and how is this transformed into a practical methodology? A question is whether centrifuge type testing could perhaps be used for reducing uncertainties related to soil–structure interaction effects.

Erosion around piles subject to fluid flow, the so-called scour phenomenon, continuously changes the soil model. A major contribution to scour occurs during extreme events such as storms, and stochastic models will be needed for predicting scour and its development patterns. An unanswered question is whether it might be possible to predict scour depths numerically, for example with the aid of CFD simulations.

8.3 How Can We Design Efficient Offshore Turbines and Install and Operate Them Safely and Economically?

Floating wind turbines are inherently highly dynamic systems that define multiple new topics for research. Many of these topics are related to the floater design, its dynamics and interaction with the control system when exposed to wave and current loads. Installation and operational aspects are very relevant here, including for bottom-fixed turbines, as they can contribute an amount to the total project cost that is similar to the initial cost of the turbine (Fig. 8.2).

Scientific challenge 8-3: To develop integrated methods for robust and reliable wind turbine design and optimisation for offshore application which enable exploration and understanding the implications of non-conventional design options.

8.3.1 Floater Design

Anchoring of semi-submersibles, spars and barges is achieved through catenary moorings, whereby it should be noted that turbulent wind speed fluctuations result in continuously changing mooring line tension. Especially for shallow-water application it can be a nontrivial task to design a well-performing mooring system that does not significantly restrict floater motions, and which operates with acceptable line tensioning. There are also ideas for alternative station-keeping systems, e.g. oscillating hydrofoils, thrusters or smart interconnection and control on the wind farm level, that warrant further investigations.

Fig. 8.2 The introduction of floating turbines poses even more challenges than current offshore turbines

Due to low natural periods of the floating system, there is a known instability if onshore blade pitch control is applied—this needs to be resolved by means of appropriate multivariable control strategies. Also, from an energetic perspective, the waves causing the floater motion imply additional energy added to the wind turbine system, which theoretically offers some potential for increased power generation.

A largely unexplored field exists regarding what control strategies are best suited for a floater and whether it might be possible to optimise, or even increase, energy production while simultaneously reducing structural loads and avoiding instabilities. The potential of active dampers and control using information about current environmental conditions has not been sufficiently studied, in particular.

The state of the art for floater structural analysis is based on rigid-body motion, whereas, in vessel design, elastic vibration modes of the structure are also considered. A question is, to what extent these are excited by the waves, and what are the effects when performing structural analysis on such a reduced model? Or should both motions and elasticity also be included for floating wind turbine modelling?

There are many different floater concepts that are currently studied, but to date an unresolved question is which concept is optimal for a given turbine, location and environmental condition. Current design selection and design optimisation is primarily based on experience, and there is a lack of concept-independent integrated floater design optimisation methodologies taking into account not only the platform itself but also other aspects influencing the dynamics such as the control system and wind turbine.

The dynamics of floaters during electrical grid network system faults can be nontrivial, but these effects have so far received little attention and the implications for design of the support structures and its control system need to be worked out.

Redundant design and design for ease of inspection constitute a relatively recent area of interest. It is not necessarily a research need or activity, but accessibility is key, and practical recommendations benefitting the design of offshore wind turbines would be of great value.

8.3.2 Installation and Operational Constraints

An essential question for planning offshore maintenance is about the best methods for forecasting the wave climate, and in what manner uncertainties could be incorporated into this process, for instance when assuming an unreliable meteorological weather forecast. An additional research topic is the search and evaluation of alternative installation concepts and to explore the concept of foundations designed for optimised ease of inspection. What are the possibilities and implications for operating turbines with slightly damaged support structures? How much redundancy should be used for the design of support structures and to what degree would down-regulating a turbine with some structural damage allow for continued energy production until a repair can be performed? Finally, would it be possible to estimate the remaining lifetime of a given support structure in an efficient way?

8.3.3 Wave Energy and Other Issues

One research question is whether it could be beneficial to combine a wind turbine with either a wave-power or ocean-power device. One observation is that it would only make sense when there is an added benefit for the wind turbines with regard to load reduction, as adding functionality also adds complexity. Hybridisation may be particularly relevant for floaters to increase energy yield, e.g. by using oscillating water column wave energy devices in the floater structure, tidal/current turbines below the floater, or reverse magnetostriction devices implemented into the mooring lines. Technologically, observations were that wave forecasting models are wind-induced and that wave height and wave direction should both be considered.

An additional option is combining multiple turbines on a single foundation or to use coupled mooring systems that connect floaters with each other, thereby reducing the number of anchors and the length of chain needed. An initial review suggests that these are perhaps not viable ideas due to anticipated general system response and control issues. Another possible serious issue could occur when a failure in one turbine starts having an impact on the other turbines on a shared platform. Other concepts that have still not been fully explored are vertical-axis or stall-regulated turbines and how they behave in the offshore environment.

A final potential issue is linked to icing as it could cause high ice-related loads for which no reliable models exist yet.

Chapter 9
Offshore Environmental Aspects

H.J. Lindeboom and S. Degraer

State of the art

The number of offshore wind farms (OWFs) is increasing rapidly, leading to questions about the cumulative environmental impact of such farms. The major impacts are caused by the noise produced in the building phase, the new hard substratum, the moving rotor blades, possible underwater noise in the production phase and the exclusion of fisheries. Danish, Dutch, British, German and Belgium monitoring programmes in individual wind farms have shown, in general, an increase in biodiversity due to the hard substratum, as well as changes in benthic and fish fauna due to the presence of the foundations and further away due to exclusion of fisheries and attraction or avoidance by some bird species or sea mammals. These effects can be qualified as partly negative, partly positive and partly unknown or so far unclear. In particular, the cumulative effects of the increasing number of wind farms need further attention. Negative effects of wind farms include effects of noise on sea mammals and possibly fish, avoidance behaviour by some bird species, and bird and bat collisions with turning rotor blades. If OWFs want to be real suppliers of green energy, ways to mitigate the negative effects need further research and development of innovative techniques.

Due to the spatial claims of wind farms, there may be conflicts with other sea users, like fisheries, but on the other hand offshore wind farms may be good sites to stimulate protein production in the open sea. A major aim of this topic is to explore innovations for design and operation of wind farms to mitigate environmental impacts and enhance multiple use of the offshore wind farm (Fig. 9.1).

H.J. Lindeboom (✉)
IMARES-Wageningen UR, Wageningen, The Netherlands
e-mail: han.lindeboom@wur.nl

S. Degraer
Royal Belgian Institute of Natural Sciences, Brussels, Belgium

G. van Kuik and J. Peinke (eds.), *Long-term Research Challenges in Wind Energy - A Research Agenda by the European Academy of Wind Energy*, Research Topics in Wind Energy 6,
DOI 10.1007/978-3-319-46919-5_9

Fig. 9.1 The alignment of the interests of wind farm operators and the fishery industry requires creativity

9.1 What Are the Cumulative Environmental Impacts of Increasing Numbers of OWFs and How Can Negative Effects Be Mitigated?

Short term effect research is executed nationally but a strong international approach is required for the long-term and cumulative effects. An international platform is necessary to exchange, compile and analyse all results in an integrated way, and initiate additional research if needed.

Scientific challenge 9-1: To analyse and quantify the impact of multiple wind farms on animal groups and the ecosystem, and to design wind farms with minimal environmental impact.

9.1.1 Impacts of an Increasing Amount of Wind Farms on Specific Animal Groups

Examples that need attention include the continuous operational noise; habitat change by epifouling communities on the turbines and scour protection, including subsequent changes in the surrounding; the recovery of benthos after trawling cessation; and the effects of turning rotor blades on bird populations. Scale and importance of impacts, as well as recovery time of impacted communities/species/areas, need more attention. Focus will be on the effects upon birds, sea mammals, benthos and fish (Fig. 9.2).

Fig. 9.2 Someone is watching us

9.1.2 Impact on the Ecosystem as a Whole, also in Relation to Natural Variability

The new EU legislation on the good environmental status of the marine ecosystem as indicated by descriptors like biodiversity, food webs and sea floor integrity may influence future legal boundaries for OWF developments. By means of an integrated analyses and modelling approach we intend to create instruments that can facilitate operators to rapidly react to these new ecosystem approaches and demands.

Long-term monitoring is needed to detect the strength of an effect in the long-run, e.g. the effect of trawling cessation on the benthos or the effect of noise on fish species and marine mammals during the operational phase. Long-term and large-scale analyses are thus essential and will be part of an internationally coordinated monitoring programme.

The natural variability in the marine ecosystem is high. More fundamental knowledge of temporal and spatial variability is needed to discriminate an anthropogenic (OWF) effect from the natural background variability.

9.1.3 How Can Wind Farms Be Designed Wind Farms with Minimal Environmental Impacts?

Long-term effects of underwater noise could have a negative impact on marine fauna. More research on the effects of noise levels and sources on different species connected to development of innovative noise-avoiding generator techniques is required for the development of appropriate mitigation strategies.

The turning rotor blades disturb or kill birds with possible negative impacts on species like gannets, auks or sea ducks. All effects depend upon location and size of the wind farms. By careful selection of sites and configuration of wind farms, avoidance of effects is possible. To do this, more knowledge on the cumulative impacts of OWFs is a prerequisite. The number of birds and their behaviour in new wind farms, in connection with their overall distribution and breeding behaviour, need to be established and modelled. The same holds for sea mammals, especially porpoises. By identifying the ecological and economically most suited areas for offshore wind now and in the future, conflicts with other sea users may be avoided and optimal use of the sea stimulated. To increase the protection potential of OWFs and to achieve a multifunctional use and biodiversity protection insight into the cumulative impacts on all underwater fauna is necessary. By coordinating the obligatory monitoring programmes with knowledge on marine ecosystem functioning and cause–effect relationships between different pressures and faunal groups, linked to technological developments of offshore wind turbines, important innovative improvements can be achieved.

9.2 How Can Multifunctional Use of Offshore Wind Farms Be Stimulated?

The demand for protein from the sea will increase in the coming years, and OWFs seem an ideal place to grow and harvest new food from the sea.

Scientific challenge 9-2: To establish possibilities for sustainable protein harvest and food production in OWFs.

9.2.1 Harvesting Natural Resources

Present offshore wind turbines are covered with mussels that grow fast and are of excellent quality, while at the base of the foundations and in the scour protection many edible crabs are found. By optimising the potential for these animals in OWF areas (e.g. by creating a more suitable habitat) and developing sustainable harvesting methods, OWFs may become an extra source of high-quality protein.

Around OWF foundations, increased numbers of fish are found. A major question is whether this is just aggregation or additional production. This question can be answered by examining reproduction, growth and survival rates. Depending on the outcome, sustainable harvesting of fish, e.g. by use of static gears, may be possible.

9.2.2 Active Aquaculture

If towed fishing gears remain prohibited in OWFs, these are ideal places for active aquaculture. Mussels, oysters, edible crabs and lobsters are easily farmed, while the potential for fish needs further investigation.

Growth of algae for consumption, biofuels or medical applications may also be possible. Food from OWFs may be realised by linking technical possibilities and innovative shaping both on and near the turbines (e.g. optimal scouring protection) with a sustainable production of high-quality marine protein.

Chapter 10
Wind Energy in the Electric Power System

H. Polinder, J. Peinke, O. Kramer, S. Lehnhoff, M. Sonnenschein and P.E. Sørensen

State of the art
The ongoing increase in renewable power generation causes a parallel overall decrease in conventional power generation from, in particular, fossil and nuclear power plants. Apart from providing market-based active power schedules, these power plants are crucial for offering ancillary services in order to guarantee a reliable stable power supply at any instant in time. Substituting these plants with renewable generation units requires the latter to be capable of providing these ancillary services. The state of the art is that grid codes are used to define the way wind turbines and wind farms have to behave when connected to the power system. In this way, they already incorporate basic ancillary services. However, frequency control is normally not provided as a regular reserve, because this would require reserving parts of the available wind capacity as stand-by capacity. Within R&D institutes, such control options were demonstrated and assessed for wind power plant clusters.

With the envisaged further development of the wind power generation towards concentrations of large wind power plants at favourable offshore and onshore locations, it is expected that interlinked variability will also continue to increase. This in turn will have strong impact on energy system planning in general, and specifically on requirements for balancing power generation and consumption. However, the state of the art in energy system planning is either to scale historic wind power time series or to apply a combination of high-resolution weather model wind speeds with a wind turbine power curve.

H. Polinder (✉)
Delft University of Technology, Delft, The Netherlands
e-mail: h.polinder@tudelft.nl

J. Peinke · O. Kramer · S. Lehnhoff · M. Sonnenschein
Carl von Ossietzky University Oldenburg, Oldenburg, Germany

P.E. Sørensen
Technical University Denmark, Kgs. Lyngby, Denmark

G. van Kuik and J. Peinke (eds.), *Long-term Research Challenges in Wind Energy - A Research Agenda by the European Academy of Wind Energy*, Research Topics in Wind Energy 6,
DOI 10.1007/978-3-319-46919-5_10

Fig. 10.1 Dark clouds above a coal fired power plant

Wind park operation planning and scheduling within virtual power plants requires a highly reliable wind power prediction down to the short range of seconds. This is necessary for enhanced service stability, uncertainty reduction, and reserve energy planning. Adequate prediction and modelling methods depend upon high-resolution measurements at wind power plant level. Approaching short time ranges, deterministic predictions become more and more impossible and stochastic methods are needed.

The development of the expansion of wind power is accompanied by a general discussion on new structures for the electric power system required to enable energy transition towards a sustainable energy system. Here we do not want to go into the general discussion of how to establish most efficiently such a new power system based on big shares of renewable energies. In this contribution we focus only on challenging aspects which are in direct relation to the wind power (Fig. 10.1).

10.1 How Can the Power System Be Kept Reliable and Stable?

Our modern societies are based on an electric power supply guaranteed by a reliable system fulfilling the demand. Current systems have been optimised to accommodate the demand with its fluctuations. With increasing renewable energies, the reliability of the electric power system has to face the problem of balancing not only the power consumption but also the power generation, which will require mechanisms for dimensioning and provision of ancillary services. The desired solution is not to build up a back-up system by traditional fossil power supply but rather to find ways that the renewable energies themselves will provide the necessary ancillary services so that the use of fossil power can be minimised. In order to reduce the need for ancillary services, it is essential to predict and model the wind power production as precisely as possible.

The increase in the shares of wind energy changes the production for electrical power from centralised controllable units to dispersed power production. The management of wind power is done on the background of different aspects ranging from the local wind conditions, over local grid requirements, to the financial and social issues. Moreover, one has to take into account that wind energy is often highly fluctuating, commonly denoted as intermittent nature of the wind power. A complete understanding and characterisation of these fluctuations are still missing but are important for optimising the power system. Even with all the desirable improved knowledge, a required new concept of a managing system has to be based on the background of big data, i.e. advanced data analytics which still provide an incomplete knowledge of the whole power generation and consumption process.

Stability of the electric power grid is a major issue. It is of high importance to be aware of possible stability cases and to quantify how far away a system is from a possible instability. This question involves the details of dynamic capacities as well as the dynamics of generating, consuming and distributing units. There are aspects of overloads and avalanche effects, which may cause large-scale grid failures, but there are also dynamic oscillations of phases, which may cause local troubles for the power system. The whole power system issue gets more complicated by the highly fluctuating character of wind power. Here the impact of wind energy on the grid system may be differentiated into "inherent" stability characteristics of the system and "added" stability issues by means of control (ancillary services, balancing, etc.). Therefore it is challenging to determine precisely, for a complex network, all kinds of instabilities, in particular if one has to work with the aforementioned incomplete knowledge of the system.

Scientific challenge 10-1: To determine precisely the stability of a network consisting of many sub-systems driven by sources and sinks with complex randomness like wind power.
Scientific challenge 10-2: To use wind energy as far as possible to establish a system accommodating the power demand in combination with an efficient ancillary system to guarantee reliability of the power system.

10.1.1 Wind Resource and Its Random Character

The reliability of a power system is based on the knowledge of the system itself. The more is known, the better strategies can be worked out and the more the wind power use can be optimised. Thus it is essential to improve models for wind power fluctuation and forecasting error with increased temporal and spatial resolution. In order to obtain models with enhanced credibility, it will be essential to incorporate the latest findings in wind meteorology research and envisaged wind turbine technology development trends. This includes, among other things, the downscaling of wind speeds from low-resolution weather models to the micro-scale conditions found at real wind turbine locations. It further involves the improvement of methodologies for capturing temporal and spatial correlation influences, as well as methodologies for quantifying and modelling the correlation between forecast errors with different time horizons (for further details see Chap. 2, "Wind and turbulence").

The fast fluctuations in wind power issue a challenge regarding the local grid dynamics. Jumps in power production in the range of seconds may cause transient stability problems leading to local grid oscillation. As the nature of these power fluctuations is far from being Gaussian, it is hard to predict them. In the case of Levy-like distributions, short-time fluctuations even with a magnitude of the installed capacity may become likely. It is also known that the accumulation of wind power over different units and regions does not show expected effects of averaging out the fluctuations. In contrast, Levy-like distributions with high likelihood of extreme events of short-time power fluctuations survive even for nationwide cumulated power. A power system based on big shares of wind energy must be able to deal with such extreme events. The first challenge will be to get a good statistical understanding of these extreme events, why do they occur, how often and with which magnitude are they expected. This leads to the question, how does the intermittency and randomness of the wind influence the stability and reliability of the power supply?

10.1.2 Concepts of Stabilisation

The intermittent nature of wind energy will make the task of grid stabilisation more challenging in different ways. Local grid structures, responsible for handling fast transients, will suffer from a faster aging of its components. Voltage, frequency and load angle oscillations may become more violent. Primary reserve as an automated reestablishment of the balance should guarantee grid stability within the range of seconds. But as more and more traditional power plants with synchronous generators are replaced, automatic stabilising primary reserves decrease.

To predict stability risks, a good understanding of the nonlinear power system in the range of seconds has to be worked out. Good stability criteria in particular for intermittent wind power sources have to be set up. For the future it will be of great importance to determine practical and cost-efficient solutions well adapted to wind energy. The wind turbines should be seen not merely as trouble-causing but also as trouble-solving elements for which an adequate management and control system has to be established.

10.1.3 Big Data and Its Information

Approaches to get a better understanding of the wind power system have to face a multi-scale problem, which becomes particularly challenging as the nature of wind power is changing with scales. Whereas large-scale wind structures are dominated by nonlinear structural dynamics, small-scale wind structures are dominated by turbulent stochastic fluctuations, for which a deterministic description seems to become impossible. The understanding of the whole power system also requires the knowledge of the grid including all other components of the grid. Another way to express this challenge is the problem of "big data" and data mining. Even if one gets all the required information, the problem remains of how to deal with the big data problem of huge amounts of measurements, predictions and how to take advantage of this detailed information. These challenges comprise the development of online architectures that are able to collect and process the data in real time and the development of efficient algorithms for computing predictions and further analyses like outlier detection and repair of incomplete time series.

10.1.4 Ancillary Services and Power Balancing with Wind Power

The most basic ancillary services include frequency control and voltage control, but also power system inertia, short-circuit power, power system oscillation damping and synchronising power are required depending on specific system design. Most of these services are already provided "naturally" through operating fossil and other power plants. These plants are connected to the grid via synchronous generators and operated below their rated power level in order to allow upward and downward regulation for matching the generation and the demand of electric power.

In this respect, there is a long-term need for dedicated R&D effort, which should explore two main directions. The first direction shows how wind can contribute to the ancillary services, either by utilising the "built-in" wind turbine control flexibility or by employing auxiliary wind plant equipment. The second direction is towards development of new ways to ensure the stability, for instance smart protection systems with less need for short-circuit power. Energy storage will probably contribute to these ancillary services and power balancing. Therefore, using the energy stored in wind farms and integrating storage utilities into wind farms need to be investigated further.

For reacting to unpredictable fluctuations in the behaviour of single turbine units as well as (correlated) sets of units, real-time reconfiguration schemes are necessary. These must be capable of guaranteeing the reliability of the operational provision of scheduled real power and ancillary services. A large number of turbine units as well as complex owner-ship structures, potentially yielding various possible conflicting target functions, require the use of distributed operating schemes. Such multi-agent

Fig. 10.2 Large scale wind energy will be fully integrated in future electricity infrastructure

systems and/or self-organisation methods must adhere to strict reliability and real-time requirements. For these tasks, corresponding management systems have to be devised in combination with the question of which amount of back-up power is needed for which time horizon (Fig. 10.2).

10.2 To Which Extend Do New Grid Structures Become Necessary?

The stability of the electric power grid is paramount. A further consequence of the increase in wind energy and the increased volatility of the energy flow makes it necessary to redesign the grid system. As solar energy systems are typically built up in parallel to wind energy, a new understanding of the distributing system has to be worked out. In particular, one has to deal with the problem that the locations of high power generation and high power consumption will change with weather conditions, sometimes even reversing the whole flow pattern. Thus, the whole power network becomes more and more dynamic for aspects of power balance issues. It is a challenge to determine the most efficient and stable solutions. Reports clearly point out that, for such new requirements, storage systems as well as a reinforced power grid become necessary. The effect of adding further links to a complex network is a current research topic, as non-intuitive and unexpected effects may happen through such additional links.

All power grid management schemes have to take into account the given grid structures, which are formed by their historical development and local conditions. In addition there is not only one grid structure but a multitude of different ones. Different voltage layers combine local grids, sometimes with weak structures and different amounts of renewable energies, in an energy system that is often international. Owners with different strategies have to come together, and local strategies to smooth out the unsteadiness of the renewable power sources may be considered. The need to transport electric energies over larger distances leads to the discussion of new high-voltage DC (HVDC) transmissions.

Scientific challenge 10-3: To optimise complex nonlinear networks for many different parameters and cases and to determine how to combine them.

10.2.1 Grid Integration of Wind Energy

Within the grid issue, the aim of which is to transform the current electric power system into one based on renewable power, wind energy is combined with the whole energy problem and will raise wind-specific challenges. With the socio-economic development of the last decades, wind (and also solar) energy is integrated into the grid on all different levels and places with a wide span between small and big units. Here one has to take into account that the grid structure has been built up based on the energy mix of conventional power plants. Based on the high volatility of wind energy it is a challenge to set up an optimised generating and distributing system. An open question is where and how to compensate for the power fluctuations most efficiently and *how to avoid congestion of the power system because of the very different power flows during low and high wind speeds.* Besides the commonly discussed reinforcement of the grid, possibilities to incorporate wind energy in local sub-grid structures with intelligent managements may open up new possibilities.

10.2.2 From Wind to Power

It is important to know how a wind turbine is converting the wind power into electric power. This conversion is commonly modelled by means of the so-called IEC power curve, relating the mean wind speed to the produced power output. This IEC power curve has been proven to be very useful to estimate annual power production. Looking closer at the power curve and the whole issue of wind power integration it is obvious that many details are not covered by the standard power curve. Wind conditions are quite often changing with site and weather conditions. Thus it is still unclear which further parameters have an impact on the power production. Meteorological conditions of the atmospheric boundary like stability, shear structures, or other structures like low level jets may alter the power production. The power output of a wind turbine during short times is definitely not understood by means of an IEC power curve. For the dynamic power production there is a chain of complex conversion processes, starting around the rotor blades and continuing over the drivetrain to the electric converting units. Along this chain one has to combine the functionality of the parts with respective control systems. For a wind farm, the complex aerodynamic and electric interaction with partner wind turbines also have to be taken into account to describe the resulting power production. For grid applications it is impossible to model all these aspects of the wind power conversion in detail, but the challenge is to have an overall understanding of the resulting conversion dynamics. This is, for example, important to make use of improved and detailed wind predictions.

10.2.3 Connection for Bigger Offshore Wind Farms

A new topic is arising due to recent developments in constructing big offshore wind farms and how to design optimal offshore infrastructures. Concerning the grid structure, the state of the art for large offshore wind farms is the application of 33 kV AC networks for the infield power collection system. In a next step this voltage is brought to a higher voltage at an offshore high-voltage transformer station, before further transmitting the power to shore. For offshore power transport distances of about 75 km and more, HVDC becomes beneficial. The first wind power plants incorporating this advanced technology have already been built. Particularly for the offshore case, big production units are in development, for which a corresponding high-capacity grid has to be built up. As grid costs essentially contribute to the cost of energy, most efficient solutions have to be found for this new technology. Further investigations should clarify whether using DC for the collection system within wind farms is more efficient, in terms of both material use and energy efficiency.

Chapter 11
Societal and Economic Aspects of Wind Energy

R.W. Künneke, P.E. Morthorst and K. Skytte

State of the art

Future sustainable energy systems significantly rely on wind energy, both offshore and onshore. Designing and developing this novel part of the power supply system is not only a technical matter—it also gives rise to fundamental societal and economic challenges.

Wind energy is developing towards becoming a mainstream means of energy provision. Future wind energy deployment needs to be technically and institutionally integrated in the electricity supply system, across national boundaries and between different energy sectors. As a consequence, the organisation of markets and regulatory frameworks has to be adapted to meet the needs and requirements of wind energy, from technological as well as socio-economic perspectives. Economic incentives in the form of production support have played a key role in promoting the deployment of wind energy in Europe over the last two decades. However, there have been frequent changes in support scheme design in many countries over the last years. This illustrates the ongoing quest of policy makers regarding how to align the promising features of wind energy as a major source of sustainable energy provision with various policy objectives on the local, national and international level, as well as different interests of industry, interest groups and citizens. There is a genuine concern that support mechanisms must be designed in a cost-efficient way, so that they strengthen public trust and position wind as a competitive means of power provision. The future regulatory setup has to stimulate entrepreneurs to drive down the cost of wind energy, and to develop innovative goods or services in support of wind energy.

R.W. Künneke (✉)
Delft University of Technology, Delft, The Netherlands
e-mail: r.w.kunneke@tudelft.nl

P.E. Morthorst · K. Skytte
Technical University Denmark, Kgs. Lyngby, Denmark

G. van Kuik and J. Peinke (eds.), *Long-term Research Challenges in Wind Energy - A Research Agenda by the European Academy of Wind Energy*, Research Topics in Wind Energy 6,
DOI 10.1007/978-3-319-46919-5_11

Fig. 11.1 How does society perceive visual impact?

Social acceptance and acceptability is decisive for the future deployment and innovation of wind energy. Societal values and expectations of final customers need to be satisfied in order to capture the promising perspectives of wind energy. The main objective of this research topic is to provide a better understanding of the socio-economic requirements that contribute to the integration of wind energy in sustainable power supply systems (Fig. 11.1).

11.1 Which Support Mechanisms Are Best Suited for Promoting Wind Energy?

The development of wind energy is strongly driven by economic considerations and EU climate goals. Offshore wind in particular has been recently supported by various financial and regulatory arrangements in order to overcome some initial economic hurdles. But ultimately wind energy needs to become a competitive means of power production. With increasing penetration rates of onshore and offshore wind, economic incentives need to be adapted, in particular to foster next generation technologies, such as offshore wind located further from shore, or airborne wind technologies. However, continuous adaptations and changes of design and level of support to wind energy can jeopardise investor confidence and, as a consequence, impact the rate of deployment of wind energy throughout Europe. A balance has to be found by policy makers and industry. An institutional design of efficient and effective support mechanisms for wind energy is necessary.

Scientific challenge 11-1: Designing regulatory support mechanisms and economic incentives that are best suited for the promotion of wind in the future European energy system.

11.1.1 Support Mechanisms

When designing economic incentives through policy support, several aspects have to be considered. There will have to be a trade-off between preventing overcompensation and wind-fall profits, stimulating innovation, easing grid integration, and exposing electricity generated from wind energy to market signals. In line with the internal market for electricity in EU, and to ensure that wind in the long run can compete with other energy sources, there is a need for addressing the important and urgent issue of improving current support policies for wind energy through competitive market measures. An example is the development of innovative offshore wind parks between several countries with devices like meshed "smart" networks or novel storage facilities. This requires a careful allocation of property and decision rights. Which market access and pricing rules apply? Under what conditions are third parties entitled to use the meshed networks and storage facilities, if at all? Who owns these facilities and is taking the financial risks? This last point is strongly related to the second subtopic.

11.1.2 Entrepreneurship and Innovation

There is an increasing need for adapting market regulation and stimulating entrepreneurship and innovation for this specific sustainable energy technology. Ultimately investments in wind energy by private wind park developers and even public network operators need to have a positive business case. These business cases are strongly affected by the regulation of the electricity sector, the cost of fossil fuel supply, and also by the matching of increasingly intermittent supply and demand.

There is a need to develop different qualitative and quantitative models to analyse and test different regulatory schemes and market conditions that stimulate entrepreneurship and innovation.

11.2 How Can System Integration of Wind Energy Be Facilitated?

Future wind energy deployment needs to be technically and institutionally integrated, across national boundaries and between different energy sectors (electricity, heat/cooling, gas, and transport). An effective analysis, design and control of the interactions and interdependencies between energy sources over time is needed, with a focus on various technical and economic parameters of energy systems. Research includes the development and improvement of energy system models as well as partial models. This is instrumental for investigating different future scenarios. From the results of these models and simulations, recommendations for priority system development, changes in the regulatory framework, and policy making can be derived. Further analyses regarding economic and social aspects of wind integration depend heavily on the results of such "energy system integration" studies.

Scientific challenge 11-2: To facilitate the economic and technical integration of variable wind energy by an intensified interaction between coupled energy markets, supported by coherent regulatory frameworks, in order to ensure stable, sustainable and cost-efficient energy systems.

11.2.1 Integration in the Electricity Supply System

The economic and technical integration of wind power depends heavily on the characteristics of the electricity system where wind power is to be integrated. The prevailing technology (e.g. thermal or hydropower generation), the availability of other renewable resources, and the wind penetration set the framework on the power supply side. On the demand side, potentially flexible loads such as industry as well as heat pumps and electric vehicles for households can support the integration of high shares of wind power. Between the two, extensions of the transmission network can also facilitate wind power integration considerably.

11.2.2 Interplay with Other Parts of the Energy System

The interplay with other parts of the energy system—e.g. existing natural gas networks, interactions with the district heating/cooling system, individual heating, or potential power storages—has an important role in the successful integration of wind energy. This interplay with other parts of the energy system is expected to be decisive to fully capture the potential for large-scale wind integration. Important aspects in this respect are innovative opportunities for energy storage.

11.3 How Can We Develop Socially Responsible Innovation in Wind Energy?

Societal acceptance for the large-scale development of wind energy is a significant challenge, sometimes even more than its technical realisation. The lack of stakeholder acceptance is a current concern. Public engagement strategies cover the general political acceptance of wind energy (on an EU-wide, national and regional level) as well as the local acceptance within a community. Both levels are interlinked. Public engagement needs to be improved by using effective participatory approaches. Future research initiatives need to go even further than an ex post analysis of interests and positions of various stakeholders. Moral and social values must be considered ex ante, i.e. in the design phase even before actually constructing these energy systems. A value-sensitive design is needed to safeguard the long-term societal acceptance and acceptability of wind energy systems (Fig. 11.2).

Scientific challenge 11-3: To develop value-sensitive designs of wind energy systems in order to safeguard long-term societal acceptance and acceptability.

Fig. 11.2 Example of a community based wind turbine in a German village

11.3.1 Onshore Wind

Today, onshore wind energy development often faces challenges on the local level. In many advanced wind markets, land use is restricted and/or population densities are high, and sites with high wind resources and good economic framework conditions may not be realised. In other cases, lack of local support—either at a political or community level—may delay developments or lead to exploration of other sites with less favourable conditions.

11.3.2 Offshore Wind

Responsible innovation of offshore wind raises different challenges as compared to onshore wind. Decision-making processes are fundamentally different; various actors (e.g. fishery, tourism, naval transport, energy production, regulatory and policy making authorities) are involved in different contexts and act according to different objectives, preferences and convictions.

Closing Remarks

The material presented here describes the research questions that, once solved, will certainly accelerate the deployment of wind energy. It shows a variety of research topics in physics, technology, design, and environmental and societal sciences. We hope that this document will initiate new discussions about the future of wind energy and its scientific challenges. The material we collected will hopefully help national and international organisations in their planning and funding of research.

Hanse-Wissenschaftskolleg
Institute for Advanced Study

Hanse-Wissenschaftskolleg, Institute for Advanced Study

The Hanse-Wissenschaftskolleg (HWK) is a foundation of the federal states of Bremen and Lower Saxony and the city of Delmenhorst. As an independent *Institute for Advanced Study,* the HWK promotes the disciplinary and interdisciplinary collaboration of internationally renowned scientists and young investigators by offering guest scholars (Fellows) the opportunity to concentrate on research projects for a certain space of time without the distractions of their regular academic responsibilities.

The **Fellows** collaborate with colleagues all over the world as well as with the local universities and research institutes. Because of the strong, on-going cooperation between the HWK and the neighboring research institutes and universities, the Fellows can make use of laboratory space for their research; this means that the HWK Fellowship can be used to realize both theoretical and experimental research projects.

The environment at the HWK is designed to foster **interdisciplinary and transdisciplinary thinking**. The *art in progress* and *Fiction Meets Science*

G. van Kuik and J. Peinke (eds.), *Long-term Research Challenges in Wind Energy - A Research Agenda by the European Academy of Wind Energy*, Research Topics in Wind Energy 6,
DOI 10.1007/978-3-319-46919-5

programs expand that thinking to the humanities and arts, generating dialog between scientists, artists, and fiction writers.

Through its **conference programs and public lecture series**, the HWK also serves as a place of intellectual debate. These conferences and workshops are distinguished by their international orientation and the interdisciplinary composition of the participants.

Co-founded by an international and culturally diverse community of scientists, embedded in the excellent academic landscape of the Northwest, and internationally linked with academic institutions and universities, the Hanse-Wissenschaftskolleg has a special atmosphere that creates optimal working and living conditions for its guest scholars.

Four research areas make up the special profile of the HWK:

Brain: The world in the brain, Brains in the world
Earth: Understanding the Anthropocene
Energy: Sustainability, Climate Protection, Reliability – Challenges for the Energy Industry, Politics, and Society
Society: Contested (Dis)Arrangements - Societies between Conflict and Cohesion

Focusing on these, the HWK strengthens and interconnects nationally and internationally recognized research and excellence clusters including those of regional universities and research institutes.

Homepage: www.h-w-k.de.